11/84   7⁹⁵

PLANETARY COMMISSION FOR GLOBAL HEALING

The
Planetary
Commission

by John Randolph Price

A Quartus Book

Library of Congress Catalog Card Number 84-61264
International Standard Book Number 0-942082-05-2

Published and distributed by
The Quartus Foundation for Spiritual Research, Inc.
P.O. Box 26683
Austin, Texas 78755

Printed in the United States of America
Planetary Commission Logo designed and produced by
Sharon Wright

This book is lovingly dedicated
to the healing and harmonizing of
this planet and all forms of life hereon.

ACKNOWLEDGEMENT

I wish to express my deepest gratitude and appreciation to the participants in the Quartus Bonds of Light Program. Your love, faith and action represented the expression of Spirit, and made possible the publishing of *The Planetary Commission* in Divine Order. Because of you, this book is truly a cooperative and exclusive activity of the membership, and will carry the magnificent Love-Light of the Quartus Society throughout the world.

CONTENTS

INTRODUCTION

When I wrote and published my first spiritual book, I was not prepared for the events and experiences that were to follow. Jan, my wife and partner for more than 30 years, had been deeply involved with me in spiritual research and the application of Truth principles, but we didn't know that the search for, and the discovery of, those evolved souls we called "Superbeings" would so dramatically change our lives.

The Superbeings was published in December 1981, and the following month we formally established The Quartus Foundation to continue our spiritual research. We quickly began to link up with individuals, groups and institutions across the United States and abroad, and through these contacts we found that we had only skimmed the surface in reporting on the great spiritual awakening. As the letters poured in daily, we could see that a tremendous evolutionary shift *was* taking place in the consciousness of mankind, but the Quickening was even more widespread than we had thought. While it still represented only a small percentage of the global population, a mighty tidal wave of spiritual energy was moving across the land, and when we were once again moved out of our cozy corners to begin the seminar and workshop programs, we were literally swept up and carried into what seemed to be another dimension in learning and growing.

The idea of having group meetings based on our personal experiences and the teachings of the illumined ones had never crossed our minds, but in April of 1982, Jan received a call from Carolyn Michael, minister of the Unity Church in Huntington, Long Island. She invited us both to come to Huntington in June to give a "Superbeing" workshop—and when Jan said that we didn't do that sort of thing, Carolyn said something to the effect that "you do now." Thanks to this beautiful Soul (Jan calls her a "living, moving, walking, talking Affirmation"), a new adventure began for us. To this day we do not promote lecture tours . . . we only go where we are invited, but since those magic days in Huntington, we have been from New York to Florida, in the Deep South and throughout Texas, up and down the Midwest and across to California.

In city after city we met extraordinary people—men and women who indeed represent the New Age—mystics, healers, teachers, Teachers of teachers, Knowers and Metaphysicians of an advanced order. In fact, we discovered even more Master Lights than we did in our original research for the first book!) At the same time, membership in the Quartus Society—the outreach group of the Foundation—was growing rapidly, and the combination of the workshop encounters and the letters from members sharing their personal spiritual experiences enabled us to truly feel, as never before, the unprecedented surge toward the Light. Things were happening! The entrance ramps to the Path were packed with seekers, all moving onto the High Road of spiritual illumination, joining those who had gone before them. Images of a new Heaven and a new Earth were coming through in meditation and stirring men and women to follow the new Vision. For others, the still, small voice from within was beginning to speak in unmistakable clarity, guiding those of every religious persuasion to the shining Path ahead.

And for sheer drama and excitement in this Journey, we had only to behold the beautiful experiences that were happening to people in every walk of life.

From Florida: "I had a death certificate signed and was

clinically dead for a period of time. I was free and on this beautiful road walking towards the light when Jesus appeared before me and said 'Not yet my child, it isn't time. You have a work to do'. I sat up in bed and frightened my mother and doctor—then went to sleep for 24 hours. When I woke up I was completely healed."

From Texas: "My dead body was waiting transportation to another part of the hospital, and my death certificate had already been signed. During this process I was fully aware of being in a field of Light. A figure appeared, identifying himself as the Wayshower and telling me that I must return because of work to do. I agreed, and I must say that my coming back to life caused quite a commotion in the hospital."

In scores of similar reports, our Elder Brother appeared on the borderland to escort the souls back into the earth plane, obviously because their mission in this incarnation had not been completed. Remember that those who are open and receptive to the Truth are still in the minority in this world, and so every Seeker and Light Bearer must be counted on for the work that is to be done in the years ahead.

From Missouri: "I have been traveling around the country since June 1983, and went through what I call a 'rebirth experience'. The only way I can explain it is like moving from the bottom of a mountain to the top in one step. It is as if my old self is completely gone."

From California: "I have experienced what is known as Cosmic Consciousness. For several hours while my husband watched in awe, I appeared transformed. I experienced the most ecstatic, intense, beautiful love. I was truly one with God and the Universe. I was part of everything and everything was a part of me. The trees glowed, their green was greener; the buildings were brighter; the air was fresher; everything and everyone was perfect and beautiful. Eventually I went back to normal again, except that I became closer to God than ever before, and that closeness has never left me. My fear of death was also gone."

From Texas: ". . . without warning, not being in a state of

praying or meditating (actually I was dressing after shower-
ing), I experienced the presence of God. For about four
hours I was immersed in His Presence. The experience, as
far as I am concerned, is absolutely indescribable. However, I can tell you that those four hours are the only time
that I have actually lived; I have just existed for 71 years in
comparison to that short period. While being immersed in
His Presence, I *knew* that death is a lie, that no one has
ever died, that we are eternal. I *knew* that I had always
been and that I always will be. In other words, I experienced eternity."

Cosmic encounters such as these are happening to men,
women and children all over the world, giving us a glimpse of
Reality and stimulating the Love Vibration that will carry us
forward in our work to heal the sense of separation from God.
For many of us, however, the experiences are not as dramatic
and we tend to dismiss them as emotional jags or quirks of
imagination. Regardless of what they are, if they move us off
dead-center and spur us on to seek the Way, the Truth and
the Light, then the experiences have been worthwhile and
have served their purpose.

You see, from a third-dimensional point of view, we do not
have forever to get our acts together. The negative energy of
the collective consciousness is moving toward critical mass, so
each one of us has a role to play as part of the Planetary Commission to reverse this force field and insure a chain reaction
of self-sustaining Good! That's the primary purpose of this
book. In Part I, Chapter One, we will define the vital
mission—discussing the critical opportunity before us and
seeing what each one of us must do to usher in a New Age of
Peace on Earth and Good Will toward all. In Chapter Two
we'll talk about "The Gathering" for the Commission . . .
those Truth Seekers and Pathfinders from all of the threads of
the Golden Cord (the metaphysical factor in all religions)
who are uniting once again for a common purpose.

So that you may be an active participant in the Commission, you must understand—from your present level of
consciousness—the Divine Plan for you individually, and for

14

mankind collectively. This calls for a close examination of your personal Akashic Records, which we'll do in Chapter Three to help you understand why you are here, what lessons you have to learn, what lessons you have already mastered, what special gifts you brought in with you, and how to develop the Life Program of your Divine Plan as a co-creator with your God-Self. Christ Self (Am)

In Part II we offer *The Commission Workbook for Self-Mastery.* This is a course of study on the Divine Reality—a comprehensive series of lessons and essays covering several chapters to help each one of us expand our thinking, lift up our consciousness, and open a channel for the Light from within. As we learn and grow together, we will be doing our part to cancel out the error of the race consciousness. This world *will* be saved. The planet *will* be healed and harmonized. And as we work together in fulfilling this mission, more love, joy, peace and abundance will be outpictured in our individual lives than we ever dreamed possible.

Thank you for being a part of this exciting adventure!

the Reyes – Ojai

John Randolph Price

U.S. has been ahead now w/ '84 election "see" how far removed from God conc. US is — sad

PART I

The Mission Defined

"There is one thing stronger than all the armies of the world and that is an idea whose time has come."

— Voltaire

CHAPTER ONE

The Future is Now

The fever pitch in the evolutionary process is not something that is building for the future. Whether we realize it or not, the new Millennium is already upon us. It is not something to come, but something that already is! We have been waiting for the New Age of Light and Love — and it's here, all around us, right now! It took place with the splitting of the atom!

Think about it! In the atom we found the solar system in miniature, and in this Age of Aquarius we will explore the far reaches of the universe. We also saw the mirror of our soul — the positive, negative and neutral particles representing our consciousness, and the nucleus or center core corresponding to the indwelling Christ or spiritual nucleus within each individual. Outer space and Inner Space — all revealed in the smallest particle of matter, and then we took the great leap forward. By splitting the atom and releasing its energy, we literally penetrated the veil. We moved through the last frontier of materiality and into the new world of spirituality.

With this new awareness, understanding and knowledge of the atom and the releasing of its energy, a new vibration in consciousness encircled the planet and the door to the Aquarian Age was flung wide open. The New Age had begun! And in this New Age the true secrets of the atom will be revealed.

At the present time, however, our scientific knowledge in this area is of the kindergarten variety.

Yes, we have entered the New Age, but now we have the responsibility, the obligation, to create the *civilization* of the Aquarian Age. That's our purpose, our mission, yours and mine. That's why we are here. Now it would be perfectly logical to assume that mankind made a detour somewhere a few years back, because if we are living in the Promised Land right now, someone has a weird sense of humor. Many of us from the Age of Pisces and before did not anticipate a New World with a billion people hungry and millions dying each year through starvation. We didn't count on the chaos, conflict and confusion—the suffering and sorrow and death all over the world. We were supposed to be living in peace and harmony by now.

But obviously we didn't unload the baggage of fear, false beliefs and error patterns as we moved out of the Age of Pisces. We carried it all with us, and that was a dangerous thing to do, particularly in entering the Aquarian Age. Remember that Aquarius is the water-bearer, and water symbolizes the inspiration of Spirit, thus the Aquarian Age is the Age of Spirituality. It should be noted that an "Age"—regardless of where it falls on the zodiac, is like a space-time energy mass, ready to serve as humanity's next evolutionary cycle. In the Piscean Age, the emphasis was on the development of the individual as a personality. In the Aquarian Age, however, the stage has been set for the unveiling of the true spirituality of man and the identification of each individual as a Spiritual being. Consequently, the Aquarian Energy Field is much different from others. As the water sign, it is prepared to break up and dissolve the negative consciousness of the individual, and if necessary, like the Flood, it will wash away all outer conditions created by the race mind, the collective consciousness of man.

As we officially entered the New Age and the Aquarian Vibration with our Piscean (and other Age) baggage, all hell broke loose. World war, the dropping of atomic bombs on Hiroshima and Nagasaki, the Korean "Police Action" and the

Viet Nam War, assassinations, crumbling cities, riots, the drug nightmare, Watergate, the hostage crises, the Middle East conflict, terrorism, all mingled and mixed with lesser or greater tragedies, depending on your perspective.

So now we ask . . . is there any way to unpack the bags from the past and make a fresh start? After all, how much time do we really have left with this kind of mass consciousness? Well, you know and I know that there has never been a problem without a solution, a question without an answer, so let's take a close look at the situation and decide what we have to do.

Where do we go from here?

First of all, where are we now? We are living on a planet, a spaceship called Earth, suspended in a vast and infinite entity called space. And even though millions of people are on the Path of Love and Light and Peace, and are dedicating their lives to Truth, there are still more than 4-billion men, women and children caught in a competitive struggle just to survive. And in this struggle, each individual thinks, feels and acts; and this mental-emotional-physical activity is registered in each individual energy field, in each consciousness. And because we are all related and connected on the subjective level, every single impulse in consciousness is impressed and registered in the collective consciousness of mankind—the universal energy field referred to as the race mind.

Now we all know the power of one individual consciousness. We know that each one of us can affect matter and change spacetime itself. We do that each day—from a positive standpoint—in our meditations, prayers and spiritual treatments. We have seen evidence of incredible changes in the body as the appearance of disease is eliminated and the physical organism is restored to health—sometimes instantly. We have also seen miraculous demonstrations of prosperity, true place success, and healed relationships where time and space were condensed to manifest a present reality. But we also know what our negative thinking and false beliefs can do—how fear, self-condemnation, unforgiveness, criticism

21

and resentment toward others can alter our physical structure and manifest as lack and limitation in our lives. Through the law of attraction, we can bring to us that which we fear the most, and hostility toward our brothers will set up such a negative vibration in our individual energy field that accidents, failure, broken relationships and other shattering experiences will be drawn directly into our lives.

Now just imagine what 4-billion of us could do grouped together in an energy field of negative consciousness. We could cover the United States in a blanket of cold never before experienced in our history. We could send storms rolling over the country to erode beaches, flood farmland and cities, destroy homes and businesses, kill people and cause damages in the billions of dollars. We could pollute our atmosphere and contaminate the soil to such an extent that certain parts of the country would be declared federal disaster areas. We could level cities and towns with earthquakes. And we could stimulate suicide bombers, establish a reign of terrorism, and cause such intense distrust between nations with the greatest weapon technology that most of the world would live in a state of panic.

Yes, we could do all of the above by staging a world-wide hate-in, but we really don't have to. We can get the same results and create the same kind of havoc on this planet by simply hypnotizing man into believing that the world of illusion is his reality. Instead of relying on God as the only Cause and the only Activity, as the only Presence and the only Power, we could make man believe that materiality is his God — thus violating the First Commandment: "Thou shalt have no other Gods before me." If we can wedge the material world solidly between man and God, we can rip man apart and cause a total sense of separation between him and his God. Then he will have to worship material idols, and in this idolatry he will fight for his passions, will kill for them, and will create such a negative consciousness in planning the fighting-to-keep and the killing-to-have, that all manner of things will be attracted to him . . . things like natural disasters, wars, famine, disease. We can unleash such evil that this planet and the people thereon will gasp for life itself!

This is exactly what we have done, and when I say "we" — I am speaking of each one of us as representatives of humanity — as cells within the collective consciousness of mankind. As the Associated Press reported on January 1, 1984: "Perhaps it was those sand-filled dump trucks parked around the White House on Thanksgiving Day. Probably it was something else. The Beirut embassy bombing. KAL Flight 007. The Marine barracks explosion. The landing on Grenada. Somewhere along the way the world appeared to become a more dangerous place in 1983. The crescendo of violence reminded Americans of the fraility of coexistence. People were thinking about the unthinkable: a nuclear war was shown in prime time. Terrorists, with kamikaze zeal, took aim. A U.S. Embassy blown apart at lunchtime; 63 dead. A U.S. Marine encampment blown apart on a Sunday morning; 241 dead. It was the bloodiest year for American servicemen since Vietnam. At home, the weather slapped at the land with high surf, heavy snows, mudslides, floods and a hurricane."

To answer the question — "Where do we go from here?" — we have to know where we want to go, and what we want to be. So let's create the picture of our destination in our minds. Just imagine a world where there is no conflict — no selfish competition — only loving cooperation. Imagine a world free of pollution, free of want, free of disease, free of disaster. Imagine a world populated with smiling, laughing, happy, joyous people — all radiantly healthy, all abundantly supplied, all loved and loving. Get the *feel* of such a world — put yourself right in the middle of it. Think of yourself as filled with life and energy and vitality, with a body that is strong and vibrant. See yourself surrounded with abundance and enjoying true prosperity. See yourself doing what you love and loving what you do, unbound and free. See yourself with the capacity to embrace and love and serve every single man and woman on this planet, regardless of who they are or what they have done — and feel that unconditional love radiating from you to all — and returning from all to you.

Barbara Marx Hubbard, a founding member of the Global Futures Network, has said that "the key now to international development is our image of the future. If we see ourselves as a universal species, co-evolving with nature to a higher stage of being, so it shall be. For as we believe, so it is done to us. As we see ourselves, so we tend to become."

We can see what we want to become, but we ask ourselves—"How in the name of God do we get from here to there?" Well, that's exactly how we get there. In the name of God.

Let's remember what God's name is. It is I AM—and the omnipotent I AM is going to take us into the Promised Land IF each one of us elects to reverse our focus and shift our minds from a concentration on materiality to a concentration on spirituality. And don't think for one moment that this renewal of our minds is going to mean less of what we call "material good" in our lives. You think Solomon was prosperous? Just remember that he tapped into the Storehouse of Plenty—the very same one that the Father has already given each one of us. Also recall that the Bible promises prosperity within thy *palaces*—not your walk-up flats or your government subsidized houses. The Bible also says that the blessings of the Lord maketh *rich*, and that you shall have an *all*-sufficiency in *all* things—and that the Lord will open unto you his good treasure, and this "treasure" includes everything you could possibly desire or need spiritually, emotionally, mentally and physically. The whole Kingdom!

But the irony of all this is: we can't have the fullness of this infinite good until we stop chasing it. Then it will chase us, and will catch us, and will shower us with so many blessings that there will not be room to receive them all—until we enlarge our capacity to receive. But as long as we are concentrating on the *getting*, we are shutting off the *givingness* of the Lord, the Spirit of God, the great I AM within each one of us.

Understand that when our minds are on God within, we are actually contemplating an *ALL-SUFFICIENCY*. But when our minds are focused on this world, on lack and limitation—

whether in health, wealth or loving relationships—we are feeding into the collective consciousness of mankind the parent thought of *insufficiency*. And the race mind will always react accordingly. Based on the law of consciousness, it will create still more lack, more limitation, more conditions of insufficiency. Mankind, seeing the spiraling effect of this activity of negative consciousness, reacts through resistance, and resistance leads to hostility, which leads to scheming and manipulation to *get* at the expense of his brother, his neighbor, the other fellow-members of the planetary community. And this darkened consciousness can do nothing but create even greater conditions for confusion, conflict and chaos. And the cycle continues, on and on and on.

Can you see now that the very monetary system of this planet has become a god to mankind? And can you see how the worship of this false god has fed the collective consciousness with so much negative energy that it is ready to explode? Can you understand why it is written that the love (the proper translation in this verse is "worship") of money is the root of all evil? Notice that it does not say *some* of the world's evils. It says *ALL* evil! And the word "all" means the whole quantity, the whole amount, the total, the utmost possible. So we can trace *ALL* of mankind's problems back to a single root: the worship of an effect. When we worship an effect, what are we doing? We are giving that effect power over us. We are making it the Lord our God, and when we do this, we are giving up our divine inheritance.

The Planetary Commission

What can we do? What can *you* do? Where do we go from here? After meditating on these questions over a period of several months, Jan and I believe that a *Planetary Commission* must be established immediately to reverse the focus from materiality to spirituality.

The word "commission" in this case simply means a group of people appointed to perform a specific duty. We have been told that each individual desiring to be a part of it must make a definite commitment to do so—and the "appointment" to the commission is automatic with the commitment. All *you* have to do is make a definite and dedicated commitment in writing to renew your mind, and with love in your heart, to choose to be a part of the healing, harmonizing influence for the salvation of this planet.

The Commission is not like the Peace Movement, or the Anti-Nuclear Movement, or any other kind of movement. We are not going to protest anything, and we're not going to resist anything. We simply want to CONSENT to let God be God! In effect, we're saying: "Hey God . . . we, as representatives of mankind, do hereby give You our permission to heal the sense of separation and restore sanity to this planet. And we'll do our part by acknowledging You and only You as our Spirit, our Substance, our Supply, and our Support."

What we're really doing is volunteering to co-create with God in the implementation of the Divine Plan . . . for each one of us . . . and for all humanity.

To serve as a Light Bearer on the Commission requires no dues. There will be no organizational structure. There will be no meetings, however on December 31, 1986, men and women of Love and Light all over the world will gather in spirit and in a simultaneous action at one specified time will release so much Love, Light and Spiritual Energy into the race mind that the hypnotic spell for the majority of mankind will break up like the thawing of a frozen lake in springtime. And that will be the true beginning of the New Age!

Why December 31, 1986? Because we have been told (the communications coming from within and through our monitors in the physical world) that 1987 will be the year of the *critical mass.* Thus far in our evolution, the mass of dark energy in the race mind has been subcritical, in that the chain reaction of negative consequences has not been self-sustaining. Massive prayer during certain periods of history, such as the long and bloody religious wars of the 1500's, the Great Plague of the 17th Century, the Civil War, two world wars, the world-wide Depression, in addition to the Light released during the appearance of Spiritual Masters at periodic intervals—penetrated the darkness. This caused a break in the chain and prevented the critical mass. But for the majority of people on the planet, prayer has now been replaced by either resistance or futility, giving the mass negative energy field a very satisfying diet. And as the density of the force field increases, it will reach the point where the darkness exceeds the light, the positive balance is lost, and the negative chain reaction goes critical—and this self-sustaining action is predicted for 1987.

Our objective is obvious. Why not reverse the polarity of the force field and achieve a critical mass of positive energy? Why not insure a chain reaction of self-sustaining Good in and around and through this planet? It can be done—and it will be! We *will* achieve a critical mass of spiritual consciousness to heal the sense of separation and restore mankind to Godkind. The fact that we *can* do it has already been proved in the laboratories. One of the major universities on the West Coast took the figure of 50-million people with a spiritual consciousness as reported in *The Superbeings*, and through the use of computers and measurements of spiritual energy radiation, made this observation: If these men and women would meditate simultaneously and release their energies into the earth's magnetic field, the entire vibration of the planet would begin to change.

And if there were no massive and dedicated counter-forces to offset the benefits, war, crime, poverty, hunger, disease, and the other problems of mankind would be eliminated. Now remember that these are scientists who made this prediction—based on laboratory measurements, analysis and computation.

In order to reach the critical mass of spiritual consciousness, our objective by December 1986 is to have 500-million people on earth simply CONSENTING to a healing of this planet and to the reign of spiritual Love and Light in this world—with no less than 50-million meditating at the same time on December 31, 1986. It is our feeling that the larger group, which will serve as the main body of the Planetary Commission and represents about ten percent of the global population, will not only maintain the negative energy mass

27

in a sub-critical state, but will also begin to break up the severe intensity of the "dark pockets" — thus preparing the collective consciousness for the massive penetration of Light on the final day of 1986. What happens if we don't break up the energy mass and the negative chain reaction goes critical? Well, that doesn't mean that we're all going up in a puff of smoke — but it does mean that once started, the negative events and conditions on the planet will become more pronounced in an accelerated timetable. The tribulations, spaced out in the past and providing a sense of relief between events, will become linked together, triggering the domino effect — with one situation stimulating the next one in a related cause and effect sequence.

The end result? I do not know. But I do know that we have a critical opportunity coming up and I want to do everything I can to make the most of it. We *can* let the Kingdom come. . . we *can* let the Will of God be done in Earth as it is in Heaven . . . which means that this world can be transformed into a heaven — right now — in the 1980's and 90's. This is no fantasy. This is not science or religious fiction. This is the Main Event of our individual lives.

On January 1, 1984, we started collecting the names of those individuals who desire and commit to be a part of this healing group. Since the list will not be used for any other purpose and will not be made available to anyone, no addresses will be required. All we seek is the name and country of the participant. Each name represents an individual consciousness, an individualized energy field, and because part of that energy always remains with the name, can you imagine the spiritual vibration emanating from the names of 500-million Light Bearers? Talk about a mighty beacon of Light!

The time for the Planetary Healing Meditation on December 31, 1986, will be at noon Greenwich time, which will encompass all the time zones in the world during that 24-hour period. (See time zone page and the Healing Meditation in the Appendix.) We do not anticipate mass gatherings at particular locations. Rather, we ask that you meet with other Light Bearers in your community, or join with members of your own family, or simply be alone in your favorite meditation chair at home. Spend at least one full hour in the healing meditation—and then remain in a peaceful, joyous and loving consciousness—continuing in that Christ Vibration throughout the day.

Between now and December 1986, much work is to be done. The period leading up to 1987 will be a time of *Preparation*. Then from 1987 on, we will embark on a three-phase program of (1) *Construction* (2) *Consolidation*, and (3) *Inauguration*.

Consider that *construction* comes from Latin words meaning *building together*. Keep in mind that the massive spiritual treatment of the race mind on Healing Day will not automatically pick us all up and drop us right in the middle of the Garden. We will continue to have our free will, and will remain under the Law of Attraction, i.e. like attracts like. This means that while the Seekers and Light Bearers will have a common bond as participants in the Good Will Task Force, there may be others who will continue to resist based on the illusions of vested interests. Therefore, the continuing work of the Commission will be to reveal the New World of peace, perfection and spiritual enlightenment—and to develop a spirit of loving cooperation among *all* people, where everyone works for the common good of all. So in essence, the *Construction* phase is to replace illusion with Reality, competition with cooperation, bringing about a real unity and the concept of true synthesis.

In the *Consolidation* stage, the emphasis will be on strengthening the bond of Divine Love that began in the Construction phase, and it is during this period that mankind will begin to sense and anticipate, as never before, what may be called The Second Coming.

It should be obvious what the next phase is all about. It is interesting that the word inauguration comes from Latin *inaugurare* which means to give sanctity to a place or official person. Does that mean that Jesus himself is returning, and that this stage will usher in his reign as Lord of the world in physical form? Or does it mean the "externalization" of the Hierarchy of Spiritual Masters? Or does it mean that the sense of separation will be healed and that each one of us will awaken fully to the Truth of our being — that the Higher Self of each individual *is* the Christ? Perhaps all three happenings and experiences will occur . . . at different intervals of time.

Dr. Jay Franssen, president of the Omega Foundation, says: "All humanity is infected with the concept of evolution and although people of different backgrounds interpret the events differently, there remains a growing feeling of an impending 'something' about to break forth into the stage of human events. From everywhere . . . from religious fundamentalists to science fiction fans with their extraterrestrials, star wars and star treks . . . to the advanced thinkers of evolution . . . there is an earnest expectation of making 'contact' with a higher dimension of being. The astrologers, seers, and predicters of the future have been fervently aligning themselves through their occult practices to gain insight into this great mystery, while the more fundamental and traditional aspirants of religion have sought more subtle and conservative methods by and through which to achieve the same information. The Bible (that great Book of ALL books) itself seems to be full of both dynamic and subtle directives admonishing man to PREPARE HIMSELF for the great event of the last days, which undoubtedly make reference to the next great quantum leap forward in the evolutionary scheme of things on a cosmic scale!

"To the very literalist, the Christ is going to return in the body and person of Jesus, while from a metaphysical (spiritual) point of view, we might very well describe the return of Christ as a new ENERGY FIELD which will seed human consciousness causing it to become the Christ individualized. From the scientific and evolutionary point of view, the Second Coming could be heralded by evolved beings who grow out of the Planetary milieu."

The important thing to realize now is that the Inauguration phase will introduce a completely new evolutionary cycle for mankind . . . one of continuing true Peace on Earth and sincere Good Will toward all individuals everywhere.

Will *you* be a part of the Planetary Commission? If you say *yes*, then the healing has begun . . . and it has begun with you!

The form identifying you as a member of The Planetary Commission is included in the Appendix. Please tear it out, sign it, and return to The Quartus Foundation at your earliest convenience. Then begin to radiate the Christ Spirit you are in Truth to this world. Open your heart and let Divine Love pour out to one and all, transmuting every negative situation within the range of your consciousness. Forgive everyone, including yourself . . . forgive the past and close the door. Let there be peace in your heart, the excitement of victory in your mind, and joyous words on your lips. Turn within and seek and find and know the only Presence, the only Power, the only Cause, the only Activity of your eternal life. Be a totally open channel for the glorious expression of this infinite YOU!

And then on December 31, 1986, join with millions around the world in the healing meditation. Don't just be a spectator in this history-making event. The addition of your individual light may be just the one to alter the balance and achieve the critical mass of spirituality.

The salvation of the world *does* depend on you!

CHAPTER TWO

The Gathering

When the idea for the Planetary Commission was first introduced publicly, someone said that if the "New Thought" people were the only ones involved, we would never reach our objective of 500-million participants. I disagreed, saying that through meditation and monitoring we now estimate that more than half-a-billion such believers are on the planet at this time working in various religious groups — and that New Thought concepts are spreading more rapidly than any other spiritual teaching.

In essence, the movement represents "the gathering" of the Truth Seekers and the Pathfinders from all of the threads of the Golden Cord (the metaphysical factor in all religions) — this time united in a common purpose. Since the early part of the last century, wave after wave of Old Souls have reentered the earth plane, and millions are strategically located throughout the world, ready for the new spiritual offensive.

According to The International New Thought Alliance, "the term 'New Thought' has been applied to the metaphysical movement which began with P. P. Quimby more than a century ago. The meaning of the words in this context was given by Judge Thomas Troward, one of the great leaders in the movement. It comes from the creative law of mind or Spirit and refers to the fact that a new thought embodied in consciousness produces a new condition."

A new thought embodied in consciousness produces a new condition! While thinking about this statement one day, I felt a strong urge to go back to "the beginning" as the basis for this particular chapter. And I believe the reason was to show that the Light Bearers of this world have been spreading "new thought" since before recorded history — with the objective of *producing new conditions* in the world!

The activists in the New Thought movement today include the spiritual beings who attempted to free the earliest inhabitants of the planet from the bondage of mortal mind. They were also the participants in seeding the race mind with revelations that led to the founding of Hinduism and Judaism, and many were the messengers who spread the teachings of Lao-tse, Zoroaster, Buddha, Confucius, Jesus, and Mohammed. And, they were deeply involved in the transcendentalist movement during the first half of the nineteenth century.

Do you not intuitively feel that *you* are a part of this Good Will Task Force? Don't discount the probability just because you are experiencing challenges. If you are a "teacher" — could there be credibility to your sharing with others if you could not show the way by your example — the example of experiencing the illusion of a problem and overcoming it by revealing the Reality behind it? Many of the Lesser Avatars, in moving from the Spiritual Realm into the third-dimensional plane, chose to play such a role on the Earth-stage in order to ultimately be of greater service. And during the role-playing their memory is sealed . . . closed until the challenge is met, and then the recognition of Identity returns.

If you are a "pupil" you may have walked on the dark side for most of your life, learning the lessons involved in the awakening process. But yes . . . you, too, were a part of that glory in ages past. Your travels simply took you deeper into the far country, and the effects of your thoughts, words and deeds — caused by your sense of separation from your Spiritual Self—produced the karmic wheel you've been riding for aeons. But in the deep recesses of your mind, you also remember the Truth of your being. New Age thinking and New Thought teaching are not strange to you, and you feel a beautiful sense of fellowship with the group of World Servers who are encircling this planet with a bond of Light and Love. One day you will fully awaken and will realize who you are. In the meantime, just know that you are here for a definite purpose and with a specific mission.

As we take our journey into the past, please understand that reviewing both preliterary and recorded history would be a lifetime project. However, in this relatively short chapter, I have attempted to go back to "the beginning" to see what this *fall of man* business was really all about, and then follow countless years of consciousness evolution to show the perennial seeding of New Thought. In doing so, I've skipped over so much history and left out the names of so many Master Lights that historians will ponder the great gaps in periods of our civilization and metaphysicians will wonder at all the Missing Persons. But from this quick summary you'll see that we were created out of the Mind of God and came forth as perfect Beings of Light, how we messed ourselves up and littered the Divine Playground, why we've been paying for it ever since, and the role of the Light Bearers in the awakening process.

There was never any unforgiveness on God's part regarding our transgressions, but in living the dream, we found it almost impossible to forgive ourselves and we've been on a gigantic guilt trip since the concept of "time" began on the third-dimensional plane. Fortunately, the Nightmare Era is quickly coming to a close, and soon, very soon, we shall see "the coming of a race from sorrow free, an age of faith and justice, truth and love and liberty."

The Beginning
As told in the Scriptures, God conceived within His Mind the *Idea* of creation—of the universe, of the dry lands, and the beasts that would walk thereon, and God saw that these Ideas

in His Mind were *good*. Then God looked within His infinite Mind and saw Himself. How beautiful an Image He saw, the sum of all, the completeness of the Ideas in His Mind. Conceived in love, the Image born in Mind became the Son, the perfect Man-Idea, and the Father-Mother Mind so loved the Son that all of Mind was given to It . . . the fullness of the God-head was embodied in the Image. And the Son became conscious of Himself as the Spirit of God, for the Thinker and the Thought were One. And God saw everything that He had made, and behold, it was *very good,* and God's work was finished; but the vast universe was without form, and man was not yet a living Soul.

The Son, the very omnipresent Spirit of God, became the first Principle, the ever-living male and female Principle, He who is eternal. He is, He was, He will ever be the power to be eternally I. I AM THAT I AM. He is Jehovah who spoke to Moses from the burning bush, the Christ who spoke through Jesus, the Self-existent One. And He is and shall ever be the Higher Self of every individual throughout all time.

In the stillness of peace and the warmth of love, Jehovah God, the Christ of God, meaning the very Truth of God, conceived within Himself, within His Mind, the idea of *you* as a living Soul. He saw Himself in expression *as* you, for creation must continue, as in the beginning. He saw Himself as life, love and wisdom. Each thought is light, and with each thought, the light of expression grew brighter. A pattern of light emerged in His Mind. To this perfect pattern of Himself in expression was given beauty and understanding. And the pattern grew brighter. He saw faith, imagination, enthusiasm . . . strength, joy, authority. And the pattern grew brighter. He saw will and freedom. The light began to pulsate, to throb. All of His thoughts of Himself in expression merged into one in the creative substance of His Being. His very focus of *you* as His expression began to alter the vibration of His universal Energy Field at the point of the divine contemplation. The expression was completed within His infinite Mind and He breathed the breath of His Life into you — and you became conscious life, a living Soul. And there were an infinite number of Souls, sparks of Light, formed in His Mind, each with the breath of Life.

You were now aware of yourself . . . you are a self-conscious mind created in the image and likeness of the Son, the Christ, the very Spirit of God . . . and you know yourself to be a spiritual being, the focus of infinite Mind. As a manifestation of the Son, you behold yourself as the Christ made manifest, and your consciousness is filled with the Knowledge of Christ. That is the nature and purpose of your Soul, for you are the Lord God in expression.

And through the omnipresent Energy Field of the Living Expressions, the Spirit of God brings forth into radiant light-form all that was conceived in Mind in the beginning. You live in a spiritual body in the Garden of Eden, which symbolizes pure spiritual consciousness, and you are one with all creation.

The Appearance of Matter

Although there was no concept of "time" as we perceive it now, a view of the cosmic sequence shows that energy forms were lowered in vibration and what we consider "matter" appeared, including the dry lands, the seas, the fish of the sea, beast of the field, and the fowl of the air . . . throughout the universe. But you were not included in this initial materialization, and for aeons you remained in your Light Body of pure energy as a joyful expression of Spirit. But then a few Souls became interested in the material plane.

Some of the Beings of Light began to move upon the planets of the many solar systems, mingling with the biological species that had been evolving since the beginning cycles of creation. Desiring to walk upon and experience the physical plane, many of the Souls projected themselves into the more advanced of the animal species and greatly accelerated the biological evolution. As they moved into the physical plane, the Spirit of God asked only one thing . . . that the Souls express or bring forth into manifestation only the divine ideas, the angelic thoughts and images of His Word. But they did not heed His counsel; they did not listen to the God-Self within, and with their free will began to experiment with their own concepts of creation. The descent into the dream-world had begun! Spirit had asked the Souls to be content with His Knowledge and to think only His thoughts after Him, to bring forth the purity of creation. But the emotional nature, developed during the process of integrating with another life form, spoke of the truth that the Soul was as a God and created the temptation to use the God-power to express other than the Divine Ideals.

The Souls began to entertain concepts in their minds that became thought forms, which materialized, became visible. And with every degree of materiality, a greater sense of awareness of materiality developed. After a time, they began to identify with their bodies and with their creations in the world, and the spiritual consciousness, the true nature of the Soul, began to fade.

Now they began to create selfishly. Competition was born. Then destruction. And protection from destruction. The world began to reflect this consciousness. Plants took on bri-ars, needles. Insects formed stingers. Poison came forth in reptiles. Self-preservation became the basic instinct of ani-mals and man. Fighting began, and the concept of death, which was never a part of the Divine Plan, was born. The God-Man had become hu-man . . . an animal man.

The spiritual consciousness, the Knowledge of the Christ Presence within, was now imprisoned by the newly developed personal ego of these souls. Imprisoned and forgotten. But even behind the wall of sense consciousness, the spiritual Ego, the true nature of the Soul, continued to work, receiv-ing thoughts from Spirit and sending forth images of the greatness and grandeur of each individual's Reality. But they did not understand this, and continued to develop their car-nal minds. And the descent into the dream-world continued deeper and deeper.

Seeing what was happening, those Souls who had remained in the spiritual realm came into the material plane to awaken their brothers from the mortal sleep in which they had fallen. Though invisible, they moved among the people as spiritual Messengers, Angels of Light, seeking to influence their minds. Some saw the light and broke away from the bondage, but most could not. They were trapped in a consciousness of angelic and animal energies, and a body that was a blend of the two.

The First Mission of the Light Bearers in Physical Form

From the cosmic view of the situation, something had to be done. And so a massive wave of Light Bearers numbering in the tens of thousands descended upon this planet, manifesting a physical body patterned after the body Idea imaged in their Souls by Spirit. They moved among those whom we now know as Neanderthal—and later Cro-Magnons—to awaken the consciousness that was trapped in the animal-man. They established what we think of today as religion, with symbols and rituals, all designed to break the spell. Some of the Light Bearers became priests and built temples with paintings and music and drama to stir the imagination. But the most that could be said for all the effort was that the mental, emotional and physical evolution was stimulated, resulting in a highly increased birth rate. Death was now a part of the dream state, and so its physical counterpart—birth—also came into being. This meant that more and more Souls were entering into the energy mutation and emerging as a specimen not originally planned for the planet. And with the consciousness-link on the third dimensional plane, a peculiar race mind (collective consciousness) was being formed in the earth's magnetic field.

Following the law of attraction, the Light Bearers began gathering in groups, creating their own particular form of civilization—with technology, culture and society much more advanced and refined than we know today. But later, many of them began cohabiting with primitive man, which can be interpreted from the sixth chapter of Genesis: "And it came to pass, when men began to multiply on the face of the earth, and daughters were born to them, that the Sons of God saw the daughters of men that they were fair; and took them wives of all which they chose."

Now the fascinating aspect in this account is that the designation "Sons of God" had not been mentioned in the Bible up to this point — so obviously the reference is to another race of people! In verse 4 of the sixth chapter of Genesis we read: "There were giants in the earth in those days; and also after that, when the *Sons of God* came in unto the daughters of men, and they bare children to them, the same became mighty men which were of old, men of renown."

This inbreeding continued for thousands of years, and the consciousness of those Light Bearers dropped further into the density of materiality, with the collective mind of the planet reflecting the darkness of the descent and rapidly moving toward critical mass. Before the final catastrophe — brought on by the disintegration of spiritual values and the misuse of their powers — many fled to other parts of the world where their superior knowledge left a lasting effect. Evidence of their migration was seen in northern Spain, Egypt, Greece, Central and South America, and legends of all of these ancient civilizations refer to "gods" who taught the secrets of the heavens and the earth.

During the next 7,500 years we went through the Great Flood, the Predynastic Age and Pyramid Age in Egypt, the civilizations of the Sumerians and the Semites, the first dynasty of Ur, and countless invasions, wars and conquests. And it was during this period that traces of the former Light Bearers, the "giants" of the earth, were lost. The reason: the remnants of that superior civilization returned to the Higher Planes to join their brothers . . . there to review the lessons learned. In time, some would walk the earth again as great teachers.

The Teachers
The next *recorded* activity from the spiritual realm occurred in our history at about 2500 B.C. when the Vedas, the ancient scriptures of what is now known as the Hindu religion began coming through. The Vedas are regarded as revelations, owing their authority to no earthly individual. From these revelations came the Four Vedas, the Upanishads,

the Bhagavad-Gita, and other sacred works. The primary thrust of this Teaching: God (Brahman) is the Supreme Spirit, the one absolute, infinite, eternal, indescribable Being — and the expression or manifestation of this Being is man's Spirit or Self (Atman), which is identical with the Supreme Spirit. In the Hindu Trinity, Vishnu is the God of Love, and Krishna is considered by the Hindus to be an incarnation of Vishnu. In the Bhagavad-Gita, Krishna says: "When goodness grows weak, when evil increases, I make myself a body. In every age I come back to deliver the holy, to destroy the sin of the sinner, to establish righteousness." He also says". . . he who seeketh Me with heart resolved, he surely findeth Me, his inmost Self."

In these ancient works we find the beginning of the "Perennial Philosophy" in written form. True, the Ancient Wisdom dates back to our unknown past, before recorded history, but man's earliest known *written* spiritual teachings are the scriptures of Hinduism, the dominant religion of India. This was the first thread in the Golden Cord that would bind all religions together in spirit. As Aldous Huxley put it: ". . . happily there is the Highest Common Factor of all religions, the Perennial Philosophy, which has always and everywhere been the metaphysical system of the prophets, saints and sages."

While Hinduism was spreading in the part of the world now known as India, a patriarch named Abraham was inspired to believe and have faith in One God, a God of justice and goodness, the basic emphasis of Judaism. And in Hebrew history, Abraham stands out as not only the first Jew, but also the "father of many nations." The great impetus of Judaism took place under Moses in the 1200's B.C. He led the Hebrews out of Egypt and into the freedom of the desert where he was given the Ten Commandments and the comprehensive Mosaic Code. And when Moses died in the land of Moab, Joshua succeeded him and completed the journey into Canaan.

But even with the advent of Hinduism and Judaism, the world continued to witness destructive wars, the fall of empires, poverty, suffering, class struggles, and corruption among the ruling class. Obviously more Light was necessary, so within a relatively short period, some of the greatest minds in recorded history came into the earth plane. Lao-tse, Zoroaster, Buddha, and Confucius all incarnated within a 53 year period between 604 B.C. and 551 B.C.

Lao-tse founded the Taoist religion in China, which is based on living in harmony with the great Impersonal Power that controls the universe. The *Tao Te Ching,* the sacred book of Taoism, teaches that heaven, earth and man were created to be in harmony with each other, but man lost the way and miscreated a world of disharmony. Here was another thread in the Golden Cord of the Perennial Philosophy!

Zoroaster, a Persian prophet who founded the Zoroastrian religion, based his teaching on the one and only God—a Supreme Being of Good Thought, Beauty, Holiness, Righteousness, Perfect Health, Dominion and Immortality. Zoroaster believed in the oneness of God and man, and that prayers were the "speaking of friend to Friend." Another thread in the Golden Cord!

Buddha, the Enlightened one, was the title of Siddhartha Gautama, the spiritual teacher of the Buddhist religion. Gautama believed in universal good will expressed from a heart of love "that knows no anger, that knows no ill will." Of equal significance was his understanding that lack, limitation, disease and death are but *illusions*—not created by God, therefore not real! His Eightfold Path to freedom encompassed right belief, right aspiration, right speech, right action, right livelihood, right effort, right thought, and right meditation. And the Golden Cord grew Stronger!

Confucius is called the "first Teacher" by the Chinese. The founder of Confucianism, he believed in a Supreme Being, but he placed the emphasis in his teachings on man's relationship to one another. He believed in the practice of ethical ideas, which would help man realize and understand the preordained harmony and justice of the Universe. Another thread in the Cord!

It is interesting that so many of the Souls who came into

41

the earth plane during prehistoric times to begin the awakening process were also the core group who took the teachings of these Masters and introduced them to the masses as Light for a new age. Yet, after thousands of cycles of the sun, much darkness still covered the land.

The Master of Love now stepped forward . . . the Soul known as the perfect embodiment of Love would now walk among men as an example of the perfect Christ Man, the True Man, the Reality of Everyone. And unto a woman with purity of consciousness a child was born, a son was given, and his name was Jesus. To those who received him, he gave them power to become the sons of God by telling them that they were exactly like him, and that they would do even greater things than he would do. A few dared to believe, and became his disciples. But most sought the darkness rather than the Light, and so he gave them his body as further demonstration of his power, of *your* power, and then he returned to the spiritual plane.

But now the spiritual evolution of mankind could never be stopped, because the Living Flame of the Christ would forever burn in the hearts of all people, and etched on the walls of time for all eternity would be the truth of man's divinity. And so the Cord became a mighty bond as the fire of Christianity was lit, and today it is the most widespread religion in the world. The wind that spread the fire of Christ can be attributed in no small measure to the tens of thousands of Light Bearers who again incarnated during and immediately following the time of Jesus.

The basic teaching of Jesus was to emphasize man's divinity, but with over 300 "Christian" denominations today, many accusing the others of blasphemy, we have to ask — what happened to the original message of Jesus? It would seem that like other "religions" developed from the spiritual teachings of a Master, institutions were organized and controlled by *men* . . . and they could be no stronger or better than the people who organized and controlled them!

From about 500 A.D. to 1500 A.D., the world went through a thousand year period called The Middle Ages — an era of turbulence, invasions, and warfare which saw the break between the Western and Eastern churches and the Crusades. But again, a Master Light appeared to penetrate the darkness and reestablish a sense of hope in the race mind. In the early part of the seventh century, **Mohammed** began teaching the belief in one God, and the attainment of peace through submission to the will of God. He was the founder of the Islamic religion, and his followers are called Moslems. Mohammed — seen as a prophet of God — banned war and violence and united Arabia in a great religious movement that eventually spread throughout the Middle East and into North America, Europe and Asia. Another thread was woven into the Golden Cord!

The Revolutions

Between 1500 and 1815, four major movements took place in the Western world resulting in economic, religious, intellectual and political revolutions. Trade was established between nations as commerce was expanded throughout the world, bringing with it the establishment of new economic systems and colonial empires. And shortly after 1500, the Protestant Revolt (Reformation) began, leading to the establishment of different religious denominations, initially Lutheranism, Calvinism, and Anglicanism. The world also experienced a major breakthrough in the arts, literature and science, with Souls such as Leonardo da Vinci, Michelangelo, Rembrandt, Voltaire, Rousseau, Bacon, Shakespeare, Milton, Cervantes, Galileo, Paracelsus, Newton, and Copernicus — to name only a few — contributing to a new age of understanding. And out of the political revolution emerged democratic forms of government, the elimination of autocratic rulers and the establishment of republics — based on the desire of people to become politically independent and have a share in their governments.

As a result of these four revolutions, the world was now ready to enter into an age of Spiritual Enlightenment, and the task force of World Servers began coming in again, concentrating at first in New England America — to become known as the "Transcendentalists" — and later as Light Bearers in the New Thought Movement.

The New Wayshowers

Transcendentalism, which began in the early 1800's was a religious philosophy based on the discipline of intuition to achieve a direct relationship between the soul and God . . . to "transcend" the senses, and also the churches and organized religion, and know the Divine Reality directly. The transcendentalists believed that the churches did nothing but institutionalize spirit itself, and that all people are spiritually equal because each individual is able to communicate with God — each person has the intuitive capacity for grasping ultimate truth. Thoreau put it this way: "It is necessary not to be a Christian to appreciate the beauty and significance of the life of Christ." And Theodore Parker, also a bright light in the movement, wrote: "The problem of transcendental philosophy is no less than this, to revise the experience of mankind; to test ethics by conscience, science by reason; to try the creeds of the churches, the constitutions of the states, by the constitution of the universe."

Ralph Waldo Emerson, one of the founders of the transcendental movement, became the "modern prophet of the Truth revival" and had a tremendous influence on what we call New Thought today. Here are a few examples of his thinking:

- "There is a principle which is the basis of things, which all speech aims to say, and all action to evolve, a simple, quiet, undescribed, undescribable presence, dwelling very peacefully in us, our rightful lord: we are not to do, but to let do, not to work, but to be worked upon; and to this homage there is a consent of all thoughtful and just men in all ages and conditions."

- "From within or from behind, a light shines through us upon things and makes us aware that we are nothing, but the light is all. A man is the facade of a temple wherein all wisdom and all good abide. What we commonly call man, the eating, drinking, planting, counting man, does not, as we know him, represent himself, but misrepresents himself. Him we do not respect, but the soul, whose organ he is, would he let it appear through his actions would make our knees bend. When it breathes through his intellect, it is genius; when it breathes through his will, it is virtue; when it flows through his affection, it is love."

• "Jesus Christ belonged to the true race of prophets. He saw with open eye the mystery of the soul. Drawn by its severe harmony, ravished with its beauty, he lived in it, and had his being there. Alone in all history he estimated the greatness of man. One man was true to what is in you and me. He saw that God incarnates himself in man, and evermore goes forth anew to take possession of his world. He said, in his jubilee of sublime emotion, 'I am divine. Through me God acts, through me, speaks. Would you see God, see me, or see thee, when thou thinkest as I now think'. But what a distortion did his doctrine and memory suffer in the same, in the next, and the following ages!"

• "When we have broken with our God of tradition, and ceased to worship the God of our intellect, God fires us with His presence."

• "We are what we think about all day long."

While Emerson was establishing the philosophical foundation for New Thought, another New Englander began experimenting with mental and spiritual healing. This pioneer was Phineus P. Quimby (1802-1866), considered the "practical master" of metaphysics. Quimby was highly successful in demonstrating the application of metaphysical principles to break the illusion of sickness and reveal the reality of radiant health. One of his students was Mary Baker Patterson, later known as Mary Baker Eddy, founder of Christian Science. (Because Christian Science claims a unique revelation from Mrs. Eddy, i.e. a fixed and final teaching, the Church does not consider itself as a part of the New Thought movement.)

Emma Curtis Hopkins, once a protege of Mrs. Eddy, later established an independent metaphysical school, and her teachings greatly influenced the founders of three major New Thought churches in America today: Nona Brooks — Divine Science . . . Ernest Holmes — Religious Science . . . and Charles and Myrtle Fillmore—Unity. And contributing to the flourishing of these Centers of Truth were hundreds of illustrious writers and teachers and thousands of "students of Truth"—all working together to set the stage for the New Age. The Light Bearers were again incarnating in record numbers and taking their positions for the work to be done.

With many new independent churches, organizations, centers and schools being formed—in addition to the pioneers named above—there was a need to unite the groups and individuals in a free and open alliance. In 1917, The International New Thought Alliance was incorporated with these stated goals: "To unify all churches, centers, and schools in the New Thought field in a spiritual framework that provides for and encourages full freedom of expression and function . . . to marshall the potential strength of the many groups and focus it into coordinated power directed toward the growth of the whole . . . to give to the world the message of spiritual healing for the whole individual . . . (and) the healing of all nations."

Since then, the New Thought movement has gathered great momentum, and *the true believers* in the "inseparable oneness of God and man" and the creative law of cause and effect are now stepping forth from every religion on the face of the earth and are moving into the staging area. The Golden Cord—"the highest Common Factor of all religions, the metaphysical system of the prophets, saints and sages"— is encircling the globe, forming a Bond of Light never before experienced on this planet.

That impending "something" is indeed about to break forth. And it is *good!*

CHAPTER THREE

The Divine Plan

For you to be an *effective* member of the Planetary Commission, you should understand your role in the implementation of the Divine Plan . . . for yourself . . . and for all mankind. And with this understanding will come a great sense of purpose, a stirring to action, an activation of your will to be a part of the co-creation. Yes, the salvation of the world *does* depend on you, for you are a part of the whole, and the whole would not be complete without you!

It is suggested that you purchase a spiral-bound notebook, which will serve as the Journal of your Divine Plan. Your Journal should be separated into four sections titled: (1) My lessons to learn in this lifetime, (2) Lessons I have already learned, (3) An inventory of my special gifts and talents, and (4) My life program. Each section will represent a part of your Master Plan for this particular lifetime. Now let's talk about the basic purpose of the Plan.

To *plan* something means to conceive, to think out, to make arrangements for — and the sum total of the conceiving, the thinking, and the arranging constitutes the PLAN. Accordingly, we can see that the *Divine Plan* is the Strategy and the Blueprint for each individual man and woman, for the entire human race, and for the planet itself, as conceived by the Infinite Thinker. In essence, we're talking about the *Will of God,* and if we think of that Will as the Cosmic Urge to express infinite Good-for-all, we begin to sense the absolute magnificence of the Plan.

We only have to look at the nature of God to see the nature of the Plan. It must encompass infinite peace and harmony, breathtaking beauty, everywhere present unconditional love, overflowing abundance, radiant perfection in mind and body, a true place of total fulfillment for each individual, unlimited joy and perfect order, and the ability of each Divine Expression of this Infinite Mind to co-create with wisdom and understanding in revealing God's Kingdom on Earth.

In defining the purpose of the Plan, let's recall that we have lost our awareness, understanding and knowledge of ourselves as perfect expressions of God — and that we are literally asleep to our True Nature. So it would be only logical to say that the objective is to awaken us to our Divine Identity. And upon awakening, each individual life begins to reflect a new Reality, and as each "unit" of consciousness is reunited with the Whole, the entire world is lifted up to a spiritual dimension of love, joy and peace.

With this basic purpose in mind, we can see that the Plan must include the elimination of fear, the dissolving of false beliefs, and the erasing of karmic debts in each individual soul. Thus, the Plan can also be viewed as a Teaching Model — and since teaching implies lessons, we can understand the necessity of various learning experiences to help us clean up and clear out consciousness, so that the Light can dispel the darkness and indeed make all things new!

Your Lessons to Learn

In the implementation of the Plan for your individual life, you have the opportunity to choose, between incarnations, those experiences and conditions that will help you in your spiritual growth. Each experience is a lesson, and each lesson learned will awaken certain spiritual qualities that are a part of your true nature. This is all contained in your Book of Life.

This Book, which is your individual Akashic Record, shows your existence as a Spiritual Being, the slipping into the dream state, your miscreations and the karmic effect, the series of lessons undertaken to eliminate karma and awaken to Truth, and the record of your journey thus far. Prior to each incarnation, your Book is reviewed for you, and you choose the lessons for your next lifetime in physical form.

What are your lessons to learn in this particular incarnation? Through meditation on this question, the answers will be revealed to you from within. You can also look without at the karmic wheel of your life for the answer. Visualize a large Ferris wheel with each compartment filled with human experiences. As the giant wheel rotates, see the lowest compartment, the one closest to the ground, tip over and release certain conditions and experiences in your life, and then move on. If you will analyze your life, you will find that as your karmic wheel makes its rotation, it will usually discharge conditions and circumstances of a similar nature. For example, you may have faced financial problems, found a way to overcome them, only to be troubled with additional experiences of lack as the wheel comes back around again. Or perhaps it's relationship or health problems that seem to be cyclical in nature in your life.

Take time now to review your life — going back as far as you can remember. Make a list of the more traumatic experiences — those of a highly charged negative nature. Note the approximate year and the general circumstances involved in each one. Do not be concerned about drawing negative energy into your consciousness by bringing each experience back into your mind. They are all still there, all securely deposited in the appropriate container on your wheel, so bring them out on paper in the first section of your Journal and take a close look at them.

A little memory jogging may help you. Think on this question: "What has been the major manifest problem in my life?" Consider:

• Personal relationships
• Job and career fulfillment

- Physical well-being and health
- Financial abundance and security
- Safety and protection
- Cooperation and assistance from others
- Add other conditions and circumstances that have been expressed in your life as negative experiences

You may want to rate yourself on a scale of 1 to 10 in these areas to give you an idea of just where you are in consciousness. Again, I want you to identify what you consider to be the major *manifest* problem in your life — where you experience periodic challenges.

Now for the second part in the "life-scan" of lessons to be learned, answer this question: "What do I fear the most in my life?" Is it the fear of being lonely? A fear of failure? A fear of lack? A fear of disease? Name your fears and write them down, and the objective in doing this is to see if your fears tie in with the major manifest problems in your life. It may take a bit of analysis on your part, but it will be well worth the time and effort.

For the third part of the life-scan of lessons to be learned, answer this question: "What do I consider the major flaws in my human consciousness?" You may want to consider one or more of the following:

- Selfishness
- Jealousy
- Resentment toward others
- Dishonesty
- Inertia
- Spiritual pride
- The inability to express love
- An inability to receive
- An inability to give
- Feelings of unworthiness
- Possessiveness
- Feelings of superiority *or* inferiority
- Cynicism
- Anger
- Other — (continue with your own personal evaluation of

your consciousness)

Now let's put it all together and answer this final question: "In looking at the manifest problems that keep recurring . . . in examining my major fears . . . and after taking inventory of the major flaws in my human consciousness . . . what are the lessons I must learn in this incarnation in order to move fully into spiritual consciousness?"

Let me give you an example of what one particular answer might be: "I see that my major karmic load in this incarnation has been one of relationships and the inability to relate to others in a loving and meaningful way. This ties in perfectly with my great fear of being alone, particularly in my later years in life. And as a result, I am possessive in my relationships, frequently jealous for no apparent reason, and filled with self-doubt and a lack of self-worth when the relationship falls apart."

So what is the lesson this individual has to learn? It is the lesson of Love . . . loving Self and others UNCONDITION-ALLY . . . unconditional love with no strings attached.

In this particular case, the lesson to be learned was obvious, but in many others, it's not. So when it just doesn't stand out in lights, you may have to go back and trace the root cause of the condition. For example, do the physical problems seem to always occur during or following a financial crisis? Did the string of broken relationships result in job dissatisfaction and unfulfillment in your career? Many people can track a problem back to one particular vulnerability, and in a great number of cases, it has been a problem related to financial lack. One Quartus member discovered in his lifescan that most of the crises situations during his adult years could be traced to financial insufficiency. In almost every instance relating to career unfulfillment or physical ailments, the root cause was concern about money. So he knew what he had to learn in this lifetime—once and for all: the laws of spiritual abundance.

Like the individual with the relationship problem, your list may reveal negative experiences all grouped under one heading. Perhaps it's health. If other problems seem to be only side-effects of this major challenge, then you know that you chose to realize God as your Life and the Perfection of your body in this incarnation. Or there may be a record of continuous dissatisfaction with the jobs you've had and with your career in general. This discontent could be a signal that your purpose in this lifetime is to find your true place — that Circle of Light where your Divine Plan can be expressed most effectively. Whatever the common pattern of less-than-desirable circumstances may be, now is the time to do something about it. If you wait, you will simply have to go through the same thing all over again, and you will be delaying your participation as a co-creator in fulfilling the Divine Plan.

Another point to consider: Every negative experience or pattern on your karmic wheel is directly related to your concept of your own self-worth. What you really believe about your own worthiness is externalized in your life and affairs. If you feel unworthy in any area of your life, you have added a burden to your soul in the form of a karmic debt, and it must, by law, be outpictured in your affairs. When you begin to equate your worthiness with God's worthiness, realizing that you are an Individualization of the Creator of this Universe, the debts are paid and you are free. When you know yourself, you will have mastered your lessons, so think about the idea that you are living right now. Do you see yourself as the Christ of God, God in individual expression? If not, do not waste another moment living the idea that you are just a "human being." Take the idea that God is individualized *as* you and let your thoughts flow from that state of consciousness. Let your emotional nature *feel* according to that Truth. Let your words reflect the totally unlimited Being that you are, and let your actions be based on the Truth that you are Omnipotence made Manifest! When you live the Idea that you are the Christ of God, the law will create everything in your life to reflect that Idea . . . abundance, love, bodily perfection, true place success, great joy, total peace, wisdom, faith, deep understanding, power, perfect order, enthusiasm, beauty, and the kind of livingness that was planned for you in the beginning.

Cutting the cord

What about karmic links of a negative nature to another person? These were obviously created in the past through resentment, hostility and fear, and you may have drawn these people back into your life so that you could learn the lesson of forgiveness. If you work daily and diligently with cleansing meditations, forgiving the past and asking the Spirit within to eliminate all error patterns and negative emotions, the remnants of those karmic links will be dissolved. But if you continue to plant seeds of condemnation and unforgiveness, you will be canceling out the releasing action and maintaining the status quo.

One way to cooperate with Spirit is to cut the psychic links that bind you to the other person. Just imagine that there is an energy cord (and there really is) between you and the particular individual. You are literally "linked up"—tied together—by a cord of negative energy attached to the solar plexus of both parties. You must *want* to cut that cord, and you must take the action to do it! Take an imaginary pair of scissors in your hand and cut the cord now. See it snap—feel it break. Your consciousness will instantly respond to this mental-emotional action, but you must see the cord break in your mind's eye. Once it is severed, begin to pour out all the unconditional love you can feel and send it to the other person. He/she has been released from you and has been moved out of your energy field to experience only the highest good. The attachment has been severed for the good of all concerned, and both of you are free. Now you can love unconditionally, with no strings attached.

You are not here to suffer

The Divine Plan for you does not have a special section on suffering. While the law of cause and effect is unequivocal, a lesson does not have to be a traumatic experience—unless you choose for it to be. You are not here to endure hardships as far as the Will of God is concerned. You are here to meet your challenges joyfully and lovingly by transmuting all negative energies in the experience or condition. If you are not

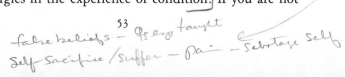

doing that, you are not fulfilling your part of the Plan. You have the power to eliminate every negative condition in your life, and you have the power to keep your life on the right track as you evolve spiritually to Christhood. Many people think that a karmic pattern of illness means that they must experience a debilitating disease, be wracked with pain, and undergo untold suffering in order to "pay for their sins" and be free of the burden. That's absolute nonsense! If you had known when you first began to experience ill health that it was nothing more than an illusion calling for your attention, simply an outpicturing of a faulty belief system and negative emotions, you could have retired to your quiet place with these kind of thoughts in your mind:

Obviously this is a signal that there is a misconception and a misunderstanding in my mind regarding the perfection of my body. I am glad that this has surfaced now so that I can eliminate these error patterns once and for all. I know that I don't have to be sick to burn away these patterns. I don't have to suffer or experience pain for the old tapes to be erased. I simply have to recognize them for what they are and rise above them in consciousness. I need only to touch the hem of the garment of my Higher Self and I am totally healed, and the permanence of the healing depends on how I choose to think and feel and act from this moment on. By being aware of the reason for my illness, I can now take the lesson and pass the test. With this under-standing, I choose to break the karmic wheel by realizing that God is my health, God is my life, and God is eternally being *me!* Therefore, it is impossible for me to suffer ill health. I can only enjoy radiant perfection and abundant well-being.

You would continue to meditate daily on your Oneness with God, forgiving everyone and everything, loving all uncondi-tionally, choosing to express only the radiant Perfection of your Christ Self, constantly accepting your wholeness in mind and body, and living each day with a sense of *having* a body fashioned after the Perfect Pattern in God-Mind. The result? Sickness would soon be so foreign to you that it would not even be a part of your consciousness. And the same would hold true for finances, personal relationships, career fulfillment, and every other aspect of your life. A "lesson" is simply a calling for forgiveness and a correction in mind.

Lessons you have already learned

Karma is the law of cause and effect, which means that it does not always have a negative connotation. In your many lives, including the present one, you have also sown good thoughts, words and deeds, and you are reaping the effect of that good work. So take an inventory of your life now to discover the lessons that you have already mastered.

You may realize that you've hardly been sick in your entire life, or that relationship problems have been few and far between, or that you have always had plenty of money to do whatever you wanted to do. Perhaps you are deeply contented with your career, and feel that you have achieved a level of success that provides fulfillment in your life. We should never look at what's wrong in our evaluations without first looking at what's right, and in doing so, we will usually find that the positive outweighs the negative.

Pause for a few minutes now and start the second section in your Divine Plan Journal. Take a review of your life and write down the beautiful highlights. From the vantage point of lessons learned, you may find that you are closer to the mountaintop than you thought. Ask yourself: What's *right* with me? What are the positives in my life? Where do I feel complete and whole?

Think about your ability to give and receive, your integrity, your evolving consciousness, your beautiful love nature . . . WHAT'S RIGHT WITH YOU?

Your Gifts and Talents

Under the law of compensation, for every lesson to be learned in a lifetime, there is also a gift to share with others — a talent for you to utilize in helping other souls master their challenges, learn their lessons, and awaken to the Truth. These gifts, talents, or characteristics of consciousness may be related in part to positive seeds sown in the past, i.e. "good" karma. And they must be used in the service of others if you are to receive the full benefit. As you work with your consciousness to rise above the challenges representing your karmic debts, if you will simultaneously use your gifts to the fullest in the service of others, the combination of the learning and the serving will so greatly accelerate your progress that you will be lifetimes ahead in your spiritual evolution. And based on where you are in consciousness, the learning-serving action in this incarnation could quickly reveal to you a literal heaven on earth!

The gifts represent the third section in your Journal, and one way to discover them is to ask yourself: "What have I always wanted to do or be? What is my heart's desire? Where are my interests? What do I enjoy doing the most?" Spend several days contemplating these questions, and the answers will begin to flow into your mind. Be sure to go back to your childhood and come forward — remembering all those yearnings you've had. Write them down and see where the common denominator is. You'll find it. It may be something that you would discount or overlook because you could not see it as a means of supporting yourself in a job or a career. Or it may be difficult for you to equate a particular heart's desire with a talent or gift. For example, if you have always had the urge to travel and see faraway lands, you may think of this as something you will do after you retire, or in your later years as strictly a leisure activity. But this may not be what Spirit has in Mind. The desire to travel may be due to the "connection" you feel for people of different backgrounds and cultures — and through the gifts of love and understanding, along with respect, admiration and a sense of brotherhood you feel, you will do your part in healing the sense of separation. Perhaps the urge to travel will be the stimulus that will cause you to investigate the travel industry, which will lead to a career opportunity, which will set the stage for future trips abroad where you will fully utilize your talents in the service of God and mankind.

Based on the above example, you can see that it will be

necessary to evaluate your heart's desire and see what the purpose behind it may be—because the gift is not always obvious. Or, you may be fully cognizant of your gift, but not know what to do with it. For greater understanding of your gift and how to use it, follow this exercise for seven days: Each morning, immediately upon awakening, ask yourself: "What do I intuitively feel my greatest strength to be? What do I intuitively feel my special gift or talent is?" As you ask each question, write down the immediate thought that comes to mind. Do not stop to analyze or evaluate the answer—just write what comes to you. Ask the two questions and write down the answers. Once that is done, ask yourself one final question: "What do I intuitively feel my Higher Self wants me to do with these special attributes?" Again, put on paper the immediate response. Continue the process each day for seven days, not missing one single day. You will find that the answers for the first three days will be colored somwhat by your ego, but by the fourth day you will have broken through to a higher realm of consciousness and the answers will pour forth with greater clarity.

What are some examples of gifts and talents? How about love, joy and the ability to see something good in everything? Also think about wisdom and understanding, and the use of these gifts to help others move through a difficult time. Consider the ability to work with children in a variety of capacities, having a sense of humor and making others laugh, and the ability to create order out of chaos through proper organization. There are musical talents, using the voice and playing various instruments. And the ability to act and create a role so real that the audience loses all sense of time and space. And what about cooking? My mother pours so much love into the preparation of food that she could make a rock taste good. Think about the ability to make something with your hands, capture beauty in a photograph, paint a picture, write a story or a poem, or teach someone—whether in a classroom or not.

There are so many additional gifts, and we have the potential to possess all of them—but certain ones are more pro-

nounced in each individual consciousness. So find your gift —
your talent — and polish it to perfection by using it to make
this a better world.

Your Life Program

Your Life Program — the fourth section in your Journal — is
all that you can desire . . . all that you can see with your
uplifted vision. The Life Program of your Divine Plan is the
abundant livingness that the Father has for you now!
Remember that the word "desire" means "from the Father" —
so all that you desire for the life more abundant — the life
more creative — the life more fulfilling — the life more loving
— the life more beautiful — the life more perfect — all of this
is Spirit knocking on the door of your consciousness saying —
"It's all yours now! All that I am, you are, and all that I have
is yours now. Take it! Love it! Enjoy it! This is my Divine Plan
for you!"

Start writing your Life Program this very day. What do you
want in life? I don't mean just making a list or a treasure
map. Lists and treasure maps are fine for the manifestation of
things, but we're talking about a Life Program of EXPERI-
ENCES now . . . and the experiences will include the things!
This is your opportunity for you to write the greatest drama
ever written — and it will be for you because it will be the
drama of *your* life. At the top of the first page write: "I see
myself . . ." and then write what you see from the standpoint
of your highest vision. The choice is totally yours . . . what
do you want to do and be and have? Drop all the can'ts, all
the inhibitions. Put yourself at the center of your world and
build around you.

As you begin to write your scenario, the first thing should
be how you see yourself spiritually. For example: "I see myself
as a spiritual being, as the very Christ of God. I am so conscious of the Christ Self within that I have become that SELF.
I live and move and have my being in Christ — as Christ — and
I am now the Master that I was created to be."

The second part of your script should be how you see the
world, and the view of yourself living in the world. "I see

myself living in a world of perfect peace and harmony . . . in a world filled with love and joy, where the sense of separation from our Source is completely healed and mankind is now living as Godkind."

Next, go back to the first section of your Journal and recall the lessons you came in to learn—and write your script showing the *mastery* of these lessons. For example, if it was the inability to enjoy a close, loving relationship with a soul mate, you might write something like this:

"I see myself in a beautiful loving relationship . . . warm and tender, yet stimulating and exciting. I see perfect unconditional love in action between the two of us, and it is so beautiful. And I love the fun, the frolic and the gaiety that I see in our relationship. We are so happy together." Write the scenes in detail . . . write the dialog between the two of you . . . describe the activities with great feeling.

Now review the second section of your Journal and take a close look at the good seeds that you have sown in the past that are now being harvested. Remember that you've learned these lessons, but you want to enhance your mastery of them in this lifetime. So if you've enjoyed good health, do not leave this attribute out of your Life Program. You may write: "I see myself filled with zest and vitality . . . I see myself with a magnificently healthy body in perfect order, where every cell is in the image of the perfect pattern, and I am whole and complete." The key is to put it into your own words . . . words that reflect your highest vision and evoke the greatest feelings of joy.

Then look at your gifts and talents and see where you can find the deepest satisfaction in your life's work—whatever that may be. We're talking about your True Place now, where you see yourself doing what you've always wanted to do. And don't say you're too old or too young or too uneducated or too whatever. Throw away the excuses. Just write scenes showing you enjoying the greatest fulfillment of your eternal life—and don't be concerned with the financial part of it, or how you're going to make money doing what you really want to do. That's a separate part of your plan, so see yourself now doing what you have always wanted to do. Just write: "I see myself . . ." and write what you see.

Once this is complete, you can phase in the abundance of supply in your Life Program by seeing yourself as wealthy as you want to be. "I see myself financially independent and totally secure with a lavish abundance!" Don't be concerned from where the money will come . . . that's none of your business . . . just see yourself overflowing with abundance and attracting bountiful prosperity from every direction, and write what you see.

Develop the other scenes of your life as you see ideal and perfect fulfillment, covering every desire, mastering every challenge, learning every lesson, capitalizing on every strength, using every gift and talent, and living life to the absolute fullest. Don't worry about your writing style, your punctuation, or perhaps your inability to create vivid and dramatic scenes. You're not writing for publication. You're writing for you! If you want to make changes in the script later, that's fine — because it's *your* Life Program. The main thing you want to do now is to set the direction of your life according to your greatest desires and your highest vision. Remember, as we see ourselves, so we tend to become!

I see myself energetic, inspired and enthusiastic! I see myself loving and loved, unconditionally! I see myself poised, confident, and filled with the power of absolute faith! I see myself as whole and complete, with an all-sufficiency of all things! I see myself with perfect judgement and as Divine Wisdom in action! I see myself as strong, mighty and power-ful! I see myself as eternal Life in perfect expression! I see myself joyous, happy and delighted to be me! I see myself enjoying the Good Will of God every single day! I see myself with perfect understanding! I see myself as the Light of the world! I see myself as God being me!

Bringing Your Life Program Into Visibility

After you have written your Life Program in detail—making sure that the scenes and visions evoke great love, great joy and excitement—go to a quiet place, sit up straight, and take several deep breaths as you focus on the Love Center in your heart. Stir up that feeling of love until you feel its magnificent vibration, then read your Life Program to your deeper mind with great feeling. Read the program either silently or aloud, whichever way stirs up the greatest emotion in you. "I see myself . . ." Read it with overflowing love. "I see myself . . ." Read it with joyful tears. "I see myself . . ." Read it with power and strength. "I see myself . . ." Read it with great happiness. Read each word with feeling, and lovingly contemplate the scenes that come into your mind. Take as long as necessary to establish, register and impress your deeper-than-conscious level of mind with the details of your Life Program.

Once you do this, you have the pattern, the mental equivalent, the mold for your Life Program—and it will remain etched in your deeper consciousness unless you change the pattern. And this is why it is wise to read your Life Program to yourself each day until the manifestation occurs.

After the pattern is established it is up to the Power, the Substance, the Creative Energy of Spirit to give it form and experience. And while Infinite Mind is going to be the primary Actor on stage now, you still have a vital role to play in the co-creation, as follows:

1. First, remember who you are. The Reality of you is pure Spirit, the very Christ of God, the Lord-God-Self that you are in Truth.

2. Secondly, understand that the Life Program you create in your mind and write in words is not your conception. It is from your Higher Self; it's how your God-Self wants to express in your world. So you can relax, knowing that you don't have to make anything happen. All you have to do is take the blueprint into consciousness to establish the pattern.

3. Thirdly, the creating-creative Energy of that Infinite Mind within you is the substance of every form and experi-

ence of your Life Program. That radiant energy—that over-flowing substance—is forever pouring, radiating from the Supermind within—right through your consciousness and out into the physical world to become all that you desire. As it moves through the pattern of your Life Program, the energy takes on the attributes of the program and begins to materialize them by changing its rate of vibration.

4. The fourth point in the co-creation activity: If you keep your focus on the *forms* of your Life Program, your mind will tell you after a time that there is not enough opportunities to meet the right mate, not enough money to accomplish your goals, not enough contacts to find your true place, not enough physical well-being to meet your objectives, not enough time to do all that you want to do, not enough wisdom to know what to do and when, not enough cooperation from others, not enough of whatever.

And the reason is because your mind will be vibrating at too low a level . . . your mental vibrations will have dropped so low that the manifestation cannot be completed. This is why you must keep your mind on the spiritual I AM within, knowing that God is the very *substance* of your Life Program and the very *activity* of your Life Program. And since there is never any lack of substance of God or the activity of God, there cannot be any obstacles to the fulfillment of your program.

In other words, there is always plenty of God! When your mind begins to understand the true meaning of *omnipresence*, it begins to vibrate to a consciousness of *plenty*, and that high vibration is the one that will bring your Life Program into visibility. You must understand that your program, which will come forth into visibility as form and experience, will be an *effect* of your consciousness. And you know that when you concentrate on the effect, you are forgetting the Cause, and when you do that, you are shutting down the power. You must look to God, Spirit, Substance alone as THE Source of your Life Program and take your mind completely off the outer world.

In essence, when you concentrate on the effects of your

world, you are lowering the vibration of your consciousness, and the lower the vibration, the more difficult it is for your good to come forth into manifestation. But when you concentrate on the Spirit-Substance of the Great I AM within you, you are raising the vibration of your consciousness. Think on this: While your mind *could* possibly conceive of a limitation of the form, it certainly could not possibly conceive of any limitation in Spirit!

Do you see now what your role in the scheme of things is? As a co-creator, you . . .

1. Take your heart's desires that Spirit has given you and write the scenario for your life according to your highest vision. The Life Program of your Divine Plan is what you really want to do and be and have in your life!

2. You give this Life Program to your consciousness with great feeling and joyful emotion so that the perfect pattern can be developed.

3. You read the program daily to protect the pattern.

4. At all other times you maintain the highest possible vibration by keeping your mind on the idea that Spirit is Omnipresent, that Spirit is appearing as your Life Program, that there is always plenty of Spirit; therefore, there can be no limitations involving your Life Program.

Soak and saturate yourself with the idea of PLENTY, and the vibration of PLENTY will manifest in your affairs . . . plenty of love, plenty of health, plenty of true place opportunities, plenty of money, plenty of time, plenty of wisdom, plenty of fun, plenty of peace, plenty of joy, plenty of vitality, plenty of inspiration, PLENTY-PLENTY-PLENTY!

The Divine Plan and True Place

By way of summarizing the concept of the Divine Plan — and to give you a broader perspective of the relationship of True Place within the Plan — let's see what Jason Andrews[1] has to say on the subject.

---

[1] Jason Andrews is the name used to identify an evolved Soul quoted at length in Chapter Three of *The Superbeings*.

Andrews: "Each man, each woman, came into physical form with a purpose, a mission. Consider people living now in America, the European countries, Russia, the Far East, throughout the world.

"Regardless of what you may call their lot in life, each was granted entrance into the manifest world to express a particular idea, faculty, a certain level of consciousness if you will.

"Never forget the uniqueness of each individual. No two are alike. Take away the mask of human identity, move past the gaze of awareness that is fastened to the illusionary world, progress on through the cave of memory and you begin to see the soft rays of a light. The closer you approach the light, the brighter is seems. This is the Light of Reality, the point where the Universal is forever becoming the Individual. It is the pressing out of God-Mind into a particular MANifestation. Here, in this secret place of the Superconsciousness, is where Uniqueness is born. God never creates identically.

"At the time of the separation when man fell away from spiritual consciousness — referring to 'time' figuratively, God etched deeply in each Soul a Divine Plan for reuniting Sonship and reestablishing the Brotherhood of Man under the Fatherhood of God. You may say that there is a Master Plan, and that each individual plan is some part of the whole, and true place symbolizes the outer expression of that inner plan. Would you not agree that true place is the special and specific activity of each individual in showing his brothers the way of At-One-Ment? Can you not see that true place is how and where you spread spiritual Light?

"The sharing of spiritual understanding is the cornerstone of true place, but included in the 'structure' is life itself: the right work, right relationships, a peaceful mind and joyful heart, and a splendid feeling of fulfillment. The Divine Plan etched in your Soul is indispensable to the whole, to the universe, for there could not be Completeness without it, without you, without each individual man and woman. ALL parts are indispensable to the whole, therefore, every Soul is equally important to God, for His Idea of the Son cannot be complete without the participating unity of all the parts.

"Can you see now that the drunk in the gutter is as important in God's eyes as the ruler of a nation? There is no such thing as a 'lost soul' — for every Soul is a Page in the Master Plan.

"True Place is how and where you are expressing the Divine Plan for your eternal life. True Place is the effect; the Plan is the cause. Pick a stranger out of the crowd. Think how special he is. Within his Soul is a link in a universal chain of Brotherhood, of Sonship, and without that particular individual, that particular link, God's perfect Idea of Himself would not be complete.

"The link represents the individual's Divine Plan. If we could go within his consciousness to read his Plan, we would see a particular gift, a special talent, for providing a useful service to others. He may not have discovered this talent as yet, but it is there. There is also a magnetic force that if followed, will lead him to the geographical location on Earth where his talent may be used most effectively. This same force will also attract the right people into his life. The Idea of dedicating his life to the love of God and his fellow man is there, along with infinite opportunities to give and share spiritual understanding through his thoughts, words and deeds. Written in the Plan is the instruction to put more into the world than he takes out . . . to be a giver rather than a taker in consciousness. He will find the Idea that his abundance is not dependent on persons or conditions, and wholeness of body is included as the perfect body Idea.

"As the Divine Plan comes forth to manifest as true place, he will find great fulfillment in his life's work, and will meet his responsibilities with joy and enthusiasm. The Golden Rule will be practiced in all his relationships, and order and harmony will be evident in every area of his life. He will feel complete, whole, total . . . spiritually, emotionally, mentally, physically. And his world will reflect this perfect balance.

"Your true place is the activity of livingness! If you cannot feel this activity taking place in your life, you must take an inventory of consciousness. To do this, look at your world. Look at your life. If it is not whole and complete, you are out of alignment with your Divine Plan. You are not on the mark. To be on the mark, you must be in your place in the universe. If you are off the mark, you are out of place.

"Also take inventory from within. What is your most compelling desire in your work life? Has there been an inner urging, a yearning of the heart for a change? Do not resist change. At the same time, use discernment, good judgement. Follow the calling but use wisdom. God has charted the course and He will lead you across the river of change at the shallow point. Or He may push you off into deep water, but will support you safely across, protecting you from the currents of race mind fears. Simply follow the inner guidance, your intuition, with confidence and peace.

"Yes, a person may feel out of place in a particular job, or in a particular community. If the discontent persists, meaning that it is not merely a mood or whim, he is being alerted to listen within. But the guidance, the instructions, must always be received in a consciousness of love. Otherwise the guidance may be misinterpreted and he may take action prematurely. He must love from the heart, pouring out the energy of love to all without exception. Love will open the divine channel of communication by dissolving the fear that had clogged it. Without a consciousness of love, a person will not find true place.

"In his consciousness of love he must follow the inner guidance to the letter, not analyzing the instructions, just following them. Take action! If he waits to evaluate and procrastinates, the entire chain of events that had been established for him will break apart. He will have missed the train, so to speak, and while another will be forthcoming, why should he wait for God's richest blessings?"

Question: What if a person does not know what he wants to do — he just knows he doesn't like what he is doing?

Andrews: "It doesn't matter what he wants to do. His God-Self knows what He wants to do, and that is all that is important. An individual's Divine Plan, expressing as true place, is a part of the WHOLE Plan to heal the separation. To reunite the Consciousness of Sonship, each man and woman must be in a position to evolve in consciousness. To evolve in consciousness, one must be in proper alignment with the universe. The Spirit within you knows precisely where this Circle of Light is, and will make the path straight before you. Even more, He will take you there!

"The Circle of Light, the Divine Plan in expression, will give you the opportunity to grow with ease. If you are outside the Circle, you will grow with adversity. The choice is yours, but you will evolve, one way or another, because the Light of At-One-Ment cannot be held back.

"Your work life within the Circle may not be a job or career as those words are commonly defined. But whatever your role, assignment, responsibility may be, it will be an opportunity to use your particular talents in the service of others. An individual may also feel that if he or she truly loves the work that fulfills a heart's desire, the compensation for that service may be below the level of financial freedom. Ask yourself: Where is that person focusing his or her attention? God is the only Source of supply, of abundance, but by looking at the job as the source, other channels will be closed. Just remember this . . . the Divine Plan for each individual includes successful service, abundant supply, ideal relationships, and radiant health, all in a divine atmosphere of peace, love, joy, fulfillment and freedom."

Question: What must we do, individually, to bring the Divine Plan for our lives into manifestation?

Andrews: "Give more where you are. Dedicate yourself to the job at hand, and meet each responsibility with joyful enthusiasm. Throw yourself into life by serving God and your fellowman to the very best of your abilities. When you give more, you love more, and the more you love, the more you give. This is working with the Law, and never forget that the activity of the Law depends on two commandments that Jesus gave us . . . 'You shall love the Lord your God with all your heart, and with all your soul, and with all your mind. This is the great and first commandment. And a second is like it. You shall love your neighbor as yourself. On these two commandments depend all the law and the prophets'.

"Has it not occurred to you what the Divine Plan is? The Divine Plan for your life *is* the Christ indwelling, your spiritual nature, your Superconsciousness, your LORD. You cannot separate Mind from the activity of Mind. The Divine Ideas representing the Plan for your life are in the thought realm of your spiritual consciousness. This is the Christ of you seeing Himself in expression! When you put on Christ through the love of Christ and your fellowman, your neighbor, you are embodying the expression. When you realize your true Identity, that Identity manifests Itself through you as your true place.

"It is impossible to be 'in place' spiritually and be out of place in your world. You cannot be spiritually rich and materially poor. You cannot be spiritually well and suffer ill health. You cannot be the Love of Spirit in expression and experience inharmonious relationships.

"The answer? The Light of God within you, your True Self, is the answer. Touch that through daily prayer and meditation. Affirm with faith and feeling that Christ in you is now made manifest in your heart, in your mind, in your body, in your affairs. Speak the word that you are *now* in your true place according to the Divine Plan that has been lovingly created especially for you.

"In a consciousness of love, work with the Law, and easily and beautifully you will be led to the Circle of Light where your Kingdom is now in expression on Earth, as it is in Heaven."

The Mission Defined

From Part I, we see that the world consciousness is quickly moving toward *critical mass*. Will the subsequent chain reaction predicted for 1987 be spiritual or mortal, positive or negative? It all depends on each one of us. By being a part of the Planetary Commission and doing whatever is necessary to change our individual consciousness, we will begin to break up some of the dark pockets of negative energy in the race mind. And through our collective efforts on December 31, 1986, we can literally turn on the Light of the World, dissolve the darkness, and begin the New Age of spirituality on Planet Earth.

Since what we are seeing today is the accumulation of our past, we have gone back into time to examine the ancient records and have traced our path from the descent into mortal consciousness to the present time. Now we can close the doors to all that is behind us, release through love and forgiveness all that has been, and co-create the future according to the Divine Plan.

From around the globe, the Light Bearers are gathering for this mission. In order for you to be a dynamic, active participant, you must understand your Master Plan for this particular incarnation. By now you have begun to recognize why you are here, and the work that you must do individually to retrain your mind and recondition your consciousness. The full scope of your Divine Plan is now coming into view, and you are eager to step out on the stage of this world as the master you were created to be.

This leads us to Part II of this book — *The Commission Workbook for Self-Mastery*. Rather than select those chapters that relate to your primary challenge, it is suggested that you first read Part II in its entirety, then start over again and spend as much time as necessary with each chapter until you feel an inner acknowledgement of understanding, a subjective comprehension of the ideas presented. Then begin to practice and live the principles in your daily life. And remember . . . no other "human" can show you a short-cut to the mountaintop. There can only be one Guide, one Teacher, one Master in your life . . . and that is the Divine Reality within you, your Spiritual Self. The writings in Part II are simply to help you rediscover your Self, and to realize that Self as the fulfillment of every desire, the answer to every need, and the solution to every problem.

PART II

The Commission Workbook for Self-Mastery

"Be not conformed to this world but be ye transformed through the renewing of your mind . . ."

—St. Paul

INTRODUCTION
TO PART II:
KNOW THYSELF

In your mortal slumber you have crossed the borderland many times in search of yourself, intuitively knowing that you are on a journey, someday to reach your destination and fully awaken to your true Identity.

In the invisible realm you did not find a heaven or hell; your consciousness remained as it was on the material plane, with only the inconvenience of space and time removed. From the teachings of the Masters, you caught a glimpse of the Self you were created to be, and you knew that your soul was in the process of unfolding, reaching out toward the ultimate moment of Christhood. Knowing that the interaction of your thought, feeling, and physical natures will accelerate your soul's evolution, you chose to incarnate again on the physical plane. And here you are today, looking for the answer, seeking the solution, searching for the magic balm that will transform sickness to health, lack to abundance, discord to harmony. You have prayed, meditated, denied and affirmed. You have imaged your good, spoken the word, and pressed onward. Some have risen above the illusion and have seen the Reality. For others, the door to the inner Kingdom was jarred, ever so slightly, and a shaft of Light penetrated into consciousness. The channel for expression was opened, only to be closed again by fear. But when only a particle of Light was released, the needed money appeared, the healing took place, the relationship was harmonized, the job was found. But why be satisfied with only temporary relief? Why be content with a morsel when you can enjoy the feast that has been prepared for you since the beginning?

Since the knowledge of yourself is the key to the Storehouse, let's pause briefly and review Who and What you are.

Remember that your true Soul is the direct expression of God *as* you. It is the Self-expression of your Higher Self, the Universal I, the Christ. In your original state, your consciousness was aware only of the Spirit within, which you recognized as the Absolute of you, the only Reality, the Creative Cause of all good. Your body was radiant substance, an invisible light form of pure energy.

When you came to dwell in the material plane and took on a physical body, a part of your Soul consciousness was lowered in vibration for the purpose of grounding and functioning on the third-dimensional plane—yet for a time there was no sense of duality, no separation in consciousness, and the wholeness of Soul functioned as the channel for the creative work of Spirit. Because the nature of Spirit is to forever remain in the Absolute, the Soul (the spiritual or Christ consciousness) was the Divine Interpreter, taking Ideas from Spirit within and interpreting those Ideas as form and experience in the material world. You lived under Grace—the Love of God in action.

But in time, the lower vibration of Soul in the dense physical body became conscious of only the third-dimensional world. From the realm of Grace you descended into the province of karma and came under the law of cause and effect. Now there was a sense of separation as the Higher Soul remained one with Spirit as the Christ Consciousness, while the lower soul dropped further into the darkness of the mortal ego.

Now you understand why you must regain your Christ-Awareness, the Soul Consciousness that you had in the beginning. The Light of the Divine Interpreter is a flame in the depths of your consciousness that can never be extinguished . . . and it can be fanned into an all-consuming fire of power and mastery if you will renew your mind, change the vibration of your energy field, and lift up your consciousness to behold the grandeur and majesty of your Soul.

One day in contemplative meditation, the thoughts that flowed through my mind spoke of the Higher Soul: "When Jesus said 'I am the Way' — it was not his personality speaking. It was the High Soul, the Christ Consciousness, that which I am. And when he said 'No one cometh unto the Father but by me' — that, too, was the voice of the Spiritual Ego. Again Jesus said, speaking from the Divine Soul, 'come unto me'. He did not say come *through* me. Once you come unto me, you come unto Spirit, for the fullness of God dwells within me. That is why it is said to 'know thyself'. I am your Self in expression, the Divine Will in expression. When you know me, you become one with me. When you become one with me, you are one with God."

Perhaps as never before, I understood the component parts of the Whole I am in Truth. I began to contemplate my physical organism and the interpenetrating energy field of my light body . . . and then my attention turned to my conscious mind . . . and then to my feeling nature and memory bank which works with the law of cause and effect in lower vibration consciousness. Higher and deeper in the inner journey I found my Soul, Spirit's idea of Itself in expression made manifest, the Mind that was in Christ Jesus, the superconsciousness of my individualized being. Entering the realm of Soul in a deep spiritual consciousness, I became aware of the Universal Christ Presence, the Absolute Self I am — and the Self that you are — THE ONE SELF! And I understood that I am you and you are me and we are one, for there is but *one* Presence. This is the unified Spirit of Sonship, the omnipresent Christ, the fathering element of all Souls throughout all creation.

Each one of us must move from the lower soul vibration into the radiant Light of the Higher Soul. This is the Kingdom Consciousness that says "No one cometh unto the Father but by me." So let's come unto this ME within us. Let's open the door and step forth into the very Presence of Christ, and there, in the omnipresence of Spirit, all sense of separation will be healed. Then we can say with true spiritual integrity, "I and the Father are one, therefore, I am one with each Soul throughout every plane of existence. The illusion of separation is over. There is only the Reality of Oneness." This is the moment of Christhood! And through this *Second Coming* in the minds and hearts of men and women throughout the world, the planet itself will be lifted up into a new dimension of wholeness.

But just an intellectual awareness of the individualized hierarchy of our be-ing will be of little benefit. We must *feel* the Truth . . . we must *know* the Truth . . . and we must *be* the Truth! That is why the first signpost on the spiritual Path reads: "KNOW THYSELF." And the reason for this admonition: we cannot expand consciousness into the *universal* until we understand and realize the true nature of the *individual*.

The remaining chapters in this book are for the purpose of retraining and renewing our minds, so that each one of us may be *all* that we were created to be. In the process, let's understand, once and for all, the principles involved in living a joyous, loving, peaceful, abundant, successful, fulfilling and free life on this planet . . . knowing that as we lift up and expand our individual consciousness, we are literally transforming the race mind.

CHAPTER FOUR

Your Personal World

1. What is included in *your* world? Start with your body — the most immediate visible form — and move out to encompass your family, your dwelling place and the environment in which you live, your job or career, your income and possessions, the people with whom you work, your friends and social activities, your community.

2. What you are seeing in this inventory of "your personal world" are *ideas* in your consciousness expressed on the third-dimensional plane. They are *your* images — and each image is nothing more nor less than your finite beliefs projected into materiality. Every person in your life is there by law of consciousness, and you are sharing mental or physical space with them through either positive or negative attraction. Even your children chose you based on the state of your consciousness.

3. Everything comes to you or is repelled from you based on the vibration of your energy field, and the vibration is established by your beliefs and convictions. Accordingly, you can see that nothing is out of place or out of order in your life. Everything is perfect based on your consciousness and the outworking of the law. Your world is a mirror of your thoughts, feelings, concepts — all pressed out in material form and experiences.

4. Do you like what you see? You are the architect and the builder, and you have designed and produced your world to the exact specifications of your consciousness. Even if you became a "health nut" to achieve a healthier body — and then ran away from your spouse, home, job, friends, and present lifestyle with the idea of starting all over again in a new city or country, in time your consciousness would create an almost exact duplication of your former world. You simply cannot run away from your world because you can't run away from yourself. You can't even escape by destroying your body, because you take your consciousness with you.

5. To run around trying to "fix" your world with the consciousness that produced the problem in the first place will only aggravate the situation even more. To change your world, you must change your consciousness. You must draw forth from within a new awareness, understanding and knowledge of the universe, the Power that sustains you, and the true nature of yourself. And with each degree in the shift in your consciousness, more Reality is revealed in your world.

6. Think of it this way: What you are experiencing in life are your finite ideas projected into materiality. However, behind what you see is what Spirit sees, and that Infinite Vision constitutes the Reality. For example, Spirit sees only a radiantly healthy body, therefore the perfect body form is the only Reality, and as your consciousness becomes more in tune with Spirit, your body will change to reflect the higher Vision.

7. The same is true for EVERYTHING in your life, in your world. The infinite Perfection forever lives behind the finite conception. Beyond the illusion is always the Reality.

8. Look at your relationships. There may be strain, turmoil and friction from your perspective, but from the Higher Vision there is only love, harmony and peace. How do you restore or harmonize a relationship? You don't have to do anything about the other person. The only person you have to do anything about is yourself. Through meditation and spiritual treatment you become one with the inner Reality and let the illusion of discord fade away.

9. What about money? In and around and through your financial affairs is the Truth of lavish abundance—the High Vision of all-sufficiency—overflowing supply to meet every need with plenty to spare and share. If you are seeing lack, limitation and insufficiency, you are looking at the illusion. But as the vibration of your consciousness becomes more spiritual—and you understand that the Spirit within you is appearing as your supply—the shadows of scarcity will dissolve.

10. Even your home and automobile are but *your* conceptions of a place to live and a means of transportation. Do they represent the expression of the Higher Consciousness? Is there beauty, ease, bountiful accommodations, harmony, and total dependability? Through your oneness with Spirit, *your* Spirit, illusory restrictions will be removed. And while the "form" of the house or car will still be finite materiality, the *experience* of joyful living and happy motoring will be Spirit appearing as the new Reality.

11. The great majority of people on the planet see only the world of illusion because they are living out of the slower vibrations and negative energies of the lower soul—the ego. What is illusion? Consider illness, suffering, lack, limitation, poverty, hunger, unemployment, conflict, crime, war, accidents, death. How can we call these seemingly real experiences illusions? Because they are not the Will of God—and only that which is the Cosmic Urge expressed is real.

12. How do we move from illusion to Reality? By awakening to our True Nature . . . by becoming of one accord with the Higher Self and *letting* that Self appear as every needed thing or experience in our lives. As you realize your oneness with the Spirit of God within, your "personal world" will change dramatically . . . your pocket of materiality on Earth will take on a new vibration, one that reflects the Higher Vision. You will have co-created a new Garden, and the Light from your Garden will be a harmonizing influence for the rest of the world.

Spiritual Activity

Make a list in your Spiritual Journal of everything you consider to be a part of "your" world. Include your body, the people around you, your place of employment—and the experiences, situations and conditions where you exert influence, and where you feel the influence of others. After your list is complete, review it with this thought in mind: *"My world is simply my consciousness projected on the screen of life, and anything I see that is not in perfect harmony is not the Will of God, therefore it is not real."*

Next, take the individual parts of your world, i.e. your body, relationships, finances, job, etc.—and contemplate each one separately. As you do, begin to see in your mind's eye and feel in your heart that God's Vision of Reality—*of Truth*—is in, around and through that particular phase of your life. See it as radiant Light filling the entire condition or situation, and know that this Activity of God is the Divine Will in Expression.

Spend several minutes at various times throughout the day contemplating that glorious Reality of total fulfillment, knowing that as you lift up your consciousness—your vision—you will be working with the Law in totally transforming your personal world.

CHAPTER FIVE

The Christ Connection

1. You are always expressing the idea of Who and What you are. If you think of yourself as a human being, you are going to experience that identity. But when you take the Idea that you are a spiritual being, that you are God individualized, and begin to *live* that Idea every moment of every day, your whole world begins to take on a different tone and shape.

2. Look at the Idea again in different words: *The Identity of God is individualized as me now. I am the Self-Expression of God. I am the Presence of God where I am. I am the Christ, Son of the Living God.*

3. This is the *Christ Idea* we're talking about — and it is this Idea that will help you to rise above the twists and turns and trials and tribulations of humanhood. Your human consciousness must begin using the Christ Idea if it is to be transformed back to its original state of perfection.

4. To conceive of and live the Idea that you are God in expression, with all the powers of God at your disposal, is to "plant your consciousness in spirituality." When you take the Christ Idea you are not just putting on a mask and playing make-believe. A mask is a disguise, a cover-up. What you are doing is revealing Reality. Remember that your whole con-

sciousness once knew itself to be God made manifest, and even when you were enveloped in sense consciousness, your Spiritual Ego, the Christ Consciousness, remained free and is right now concentrated in a Love Vibration within your energy field.

5. Another point of vital importance: Right within the darkened part of your lower-vibration mind, buried beneath layers of sense consciousness, is a *memory* . . . a faint mental impression, an *idea* of all that you once were. Now think of it. When you say "I am the son of God . . . God is as me now . . . God is expressing as me" — you are refreshing your memory. You are agreeing with something that you already know!

6. When you formulate this Master Idea of Who and What you are, the Idea first enters your intellectual awareness, then moves down into the feeling nature (subconscious). Based on the law of attraction, the Idea begins to seek out the corresponding thought in your memory bank (subconscious) and quickly attaches itself to it — and that ancient memory is stimulated and its "dawning" begins to illumine your conscious mind with new understanding.

7. When you first begin to understand Who and What you are, it is much like the Prodigal Son who "came to himself." Your memory has been stirred and the awakening process has begun. As the awakening deepens and expands in your thinking and feeling natures, a new vibration forms in consciousness. The energy of the Christ Idea begins to think and to know itself as a Divine Idea, an Idea *related* to your Higher Soul, your Superconsciousness. And so the Christ Idea says to itself, "I will arise and go to my father." Its vibration begins to move through the inner space of consciousness, through the walls and layers of error thoughts and false beliefs, through levels of hardened patterns of fear and doubt.

8. And when the Christ Idea is still far from home, the Soul of Reality within, the Christ Truth, sees it and begins moving toward it. "But when he was yet a great way off his father saw him and had compassion, and ran and fell on his neck and kissed him." This means that as the Christ Idea

begins to move through consciousness toward the Reality within, your Higher Soul begins to radiate and move toward the Divine Idea. The "kiss" is the union of human consciousness expressing as the Christ Idea, and the Christ Consciousness of the Higher Soul. This is the Realization. This is the Experience. At that glorious moment the merging Lights become one and your being is filled with the Light of Truth. You are born anew.

9. Do you see how your Identity-Idea can change your life? It changes the way you think, feel and act — and literally transforms your consciousness. When that mystical union with Self takes place, you become "Christed" — and all the Powers of God come forth in a Master Mind Consciousness and you live under the Law of Grace. But you do not have to wait for Christhood to begin experiencing harmony in your affairs. While the Christ Idea is making its journey home across the planes of inner consciousness, the spiritual law of cause and effect will be working *for* you.

10. In effect, this Law says to you: "Whatever you conceive yourself to be, I will outpicture in your life and affairs. If your mind fluctuates between Godhood and humanhood, I will reflect that vacillation in your world. However, as you *live* the Idea that *God is as me now* with great joy, love, enthusiasm and dedication — as you fill your mind and heart with this Divine Idea, I will work effortlessly to bring forth in your life all that this Idea represents. Then when the Experience comes and you begin to live under Grace, everything in your life will be a direct reflection of the Christ Truth."

11. Until the Experience, until the Realization, the Christ Idea *is* the Christ Indwelling. It represents the Divine Potential. It is the Christ Child born in the stable of your lower soul . . . "Unto us a child is born, unto us a son is given and the government shall be upon his shoulder." And the governing of your life *will* be upon his shoulder because the Christ Idea will be directing the Law, sending it before you to straighten out every crooked place!

12. ". . . and his name shall be called Wonderful, Counsellor, the mighty God, the everlasting Father, the Prince of

Peace." How can an *Idea* be given these holy attributes of God? Think about it for a moment. The Law says that whatever you hold in consciousness will be outpictured in your world. Accordingly, the Christ Idea held in consciousness will heal your body, prosper your affairs, and restore your relationships. Would you not call this Wonderful?

13. The Christ Idea symbolizes infinite Intelligence, so the Law will interpret this Knowingness for you and will direct and govern your affairs. Could you ask for a better Counsellor? "Mighty God" means Power, and the Law working through and as the Divine Idea in consciousness is omnipotent. There is nothing it cannot do for you.

14. The "everlasting Father" is representative of Divine Love; and the Law, working through this aspect of the Christ Idea, will smooth out every difficulty and attract love to you in full measure. As the "Prince of Peace" the Christ Idea will quiet your mind, still your emotions and bring forth a sense of peace that passeth all understanding. The Christ Idea is the Christ Child who will overcome the world, *your* world, and "of the increase of His government and peace there shall be no end. . ."

Spiritual Activity

Spend from ten to fifteen minutes today — in two separate sessions — contemplating the thoughts shown below. Immediately following the meditation periods, listen within and write in your Spiritual Journal the related ideas that come to you.

The Spirit of God is right where I am, and I am eternally aware of this Presence. This Spirit, my Spirit, conceived within Its Mind an Idea of Itself in expression. I am that Idea made manifest. God is expressing as me now. I am the expression of God. I am the Christ.

The Law, the Creative Energy of God-Mind, is flowing through the idea that I am now living. That Idea is the Christ, the Self-expression of God that I am, and my world becomes a reflection of that Idea.

As Christ is the Healing Principle, so the Law restores my body according to the Perfect Pattern.

As Christ is the Abundance Principle, an all-sufficiency of supply now manifests for my use.

As Christ is the Harmony Principle, all of my relationships are lovingly renewed and strengthened.

I am now the Living Truth of Wholeness and Fulfillment. And it is so!

CHAPTER SIX

No Man's Land

1. As you move through the inner space of consciousness toward the union with Self, there is a bridge you must pass over. It is the link between the third and fourth dimensions, and it is on this bridge that you shed the remaining particles of error thoughts and negative beliefs and go through the final cleansing. It has been called "no man's land" because it is the point of separating with the ego just before uniting with the Total Self in consciousness.

2. As the bridge comes into view, your world may seem to turn upside down, and the reason is because you are beginning the process of letting go of everything that seemed secure to you in the third-dimensional world. Depending on the degree of your functioning in lower mind vibrations, your ego may choose to do battle as you step on to the bridge, and it will do whatever is necessary to save itself. If that means creating an insufficiency of funds, it will do it, because this effect could very well cause you to step back into third-dimensional thinking and assume control of the situation, which would put the ego back on the throne of the mental world. Another ego tantrum may give the appearance of a business failure, or the interruption of a successful career, or perhaps a physical ailment. The ego simply wants to show who is boss.

3. Throughout the world, men and women are moving up in consciousness and are reaching the higher realms, and as they close in on the borderland of the Kingdom, the ego starts to panic. It knows that when the bridge is crossed, its role in the scheme of things will be reduced from master to servant.

4. But remember the story of the Prodigal Son. As you step out on the bridge, the Christ within, the very Spirit of God, comes forward to meet you—and this Omnipotent Presence will meet you at the half-way point. You don't have to make the journey across no man's land alone. You only have to go half-way, and at this center point you are engulfed in the Light and are taken into the Spiritual Dimension with the everlasting arms of Love around you.

5. How do you navigate the last mile as the ego begins to fight for its life? You totally surrender to God. You literally take on an "I don't care" attitude—regardless of what is going on in the world around you. You turn *everything* in your life over to the indwelling Christ and give up all concern, knowing that your God-Self is the solution to every problem and the answer to every need, and that Spirit cannot let you down because it is against God's nature to do so!

6. In the booklet *The Manifestation Process*, I point out that we know when we have truly surrendered by "the total lack of concern, anxiety and outside pressure in our consciousness. This negative energy will have been replaced with the positive vibrations of peace, joy and confidence . . ." Of course, surrendering and reaching this state of mind is sometimes easier said than done, because even as you begin to give up to the Higher Power, the ego will do what it can to bring you out on the battlefield again.

7. Total surrender means to not resist, attack, or fear anything. It means to have the courage, perhaps for the first time in your life, to put your trust in God and only God. It means to place your faith in Omnipotence and not in the potential actions of your creditors . . . to trust the Activity of Spirit and not the illusory activities of this world . . . to believe in the One Cause rather than in negative appearances.

8. If checks are bouncing, creditors are calling, your business is failing, your spouse has left, the children have turned savage, and your body seems to be falling apart, what is the worst thing that can happen? You cannot die. You are not going to be eaten alive. You cannot really lose anything because all material effects can be recreated. And the only person you can do anything about is yourself. So what are you afraid of? If you say anything other than "nothing" — it's the ego talking.

9. When you get to the end of your rope, let go. Underneath are the Everlasting Arms. God is your support! Remember the promise: "For I, Jehovah thy God, will hold thy right hand, saying unto thee, Fear not; I will help thee." God is your security! "I will fear no evil for thou art with me." When you turn away from the illusions of this world and completely surrender your life and affairs to the infinite Love, Wisdom, Power and Activity of Spirit, the ego is smothered by the blanket of spiritual Light that enters your consciousness. And you are free to complete your journey home, knowing that God will meet you half-way.

Spiritual Activity
Read and meditate on Psalm 27 and Psalm 91.

CHAPTER SEVEN

The Choice is Yours

1. Some people feel that it is spiritually wrong to desire anything—and a few metaphysical writers follow the theme of "desireless living" in their books. They are basing this concept on Jesus' instructions to "Take no thought for your life . . . your Father knoweth that ye have need of these things. But rather seek ye the kingdom of God; and all these things shall be added unto you." (Read Luke 12:22-32.)

2. To understand the meaning behind this instruction, we have to look at it from two different levels of consciousness. The first level is predominantly "human"—in that the full robe of spiritual consciousness has not been put on. On this level we work spiritually to uplift and expand consciousness, but at the same time we are given the opportunity to shape and mold our world through the use of the various power centers established within our consciousness. We have the power of free will to determine what we want in life, and we have the authority to call forth our good through the powers of decree, imagination, enthusiasm, joy and faith. And then we *release* our desires, our needs, to the higher Vision and Power of the Christ Self within, and we give no further thought to a concept of NEED.

3. As I stated in the Author's Preface to *The Superbeings,*

"The key thought that came forth from within went some-thing like this: 'Claim your good. Imagine your good. Speak the word for your good. Then care not if your good ever comes to pass'. That seemed to be quite a contradiction at first. If I desired something with all my heart, I *did* care if the desire was fulfilled or not. But the *caring*, which is another word for worry and concern, was actually diverting the power flow. I was told to choose what I wanted, see it as an actuality, call it forth into visible form and experience — then not be concerned about the outcome, regardless of how desperate the need." In other words, "TAKE NO THOUGHT!" Let it go, release it, turn it over to a Higher Power and get out of the way of the marvelous creative activity of Spirit.

4. Now let's move up to a higher level of consciousness and see what it means to "take no thought." Simply stated, when your consciousness of Truth is the ruling force in your mental and emotional natures, this consciousness will automatically be reflected or outpictured in your world and affairs without any concentrated effort (thought) on your part. This is what it means to live under grace, as a beholder of God in action through you. And this is our objective, our ultimate goal, but until we reach that level of consciousness, let's use the faculties, the powers, the attributes that we have at our dis-posal . . . one of which is the ability to *choose!*

5. Pause for a moment now and look at your life. Are you experiencing any kind of lack or limitation? Are you suffering from any type of physical ailment? Is your work boring and unfulfilling? Whether you answered "yes" or "no" to these questions is not the point; the point is that you are simply experiencing that which you have already chosen! Think of it this way: You could not have lack, sickness, unfulfillment and strained relationships unless you first chose these particu-lar experiences in your life. How can this be? You cannot have anything in life — positive or negative — unless you *accept* it, and you cannot accept it unless you make up your mind to do so, and when you make up your mind about anything, that is the action of *choosing!*

6. It should be obvious to you now that you are constantly

choosing every moment of every day, so isn't it time to start choosing rightly? Isn't today the day to start acting rather than reacting? As you are sitting and reading this book, why not make a firm decision in your mind to do what you want to do, be what you want to be, and have what you want to have. Begin now to take control of your mind and emotions, and to focus on the peace, joy, love, abundance and radiant perfection that have always been yours. *Choose this day that which you desire!* As Charles Lelly has so aptly put it, "We are the master of our own destiny only in the measure of our ability to choose wisely and constructively."

Spiritual Activity

If you have completed the Life Program section of your Spiritual Journal — as discussed in Chapter Three — go back and review the story that you have written about your life. If you have not begun this phase of your Divine Plan, you are encouraged to do so at once. Choose the experiences and activities that will be a part of your life beginning this day. Stake your claim to *all* your good — then release everything to Spirit and relax. Let go and let God be God! And remember, take no thought as to how your good is to come about. "God works in mysterious ways His wonders to perform" . . . "My ways are ingenious, my methods are sure" . . . "Trust in me, commit your ways unto me."

CHAPTER EIGHT

The All-in-All of Self

1. You always receive according to what you recognize your True Self to be. One day in meditation I was told: "That which you believe I am, I AM." In other words, whatever your consciousness attributes to your God-Self will determine your demonstration. If you believe the Spirit within to be the life and health of your body, then this perfection will be outpictured in your body. If you believe that your Master Self is your supply, you will never be at a loss for money.

2. This is the key: You must be aware that whatever you seek in life, you already have — because your Higher Self *IS* it! Whatever fulfillment you associate with your Self will be yours. So if you want to demonstrate radiant health, gain the consciousness that your True Self IS your health. If you want more financial abundance in your life, gain the consciousness that your Self IS your abundance, IS your prosperity. When you achieve a consciousness of your Inner Self as the ALL-IN-ALL, the Giver and the Gift, you will be stepping up to mastery.

3. In my experience I have found that I could pray and affirm and visualize and speak the word for days on end with nothing happening — because I was trying to *make* something happen. Only when I went within and *let* my God-Self enter

my awareness—to where I could truly say "I AM" from a sense of spiritual identity—did my world begin to change. And I have realized that it is my consciousness of God *AS* the needed thing or experience that causes the change . . . and the greater and deeper the consciousness, the more dramatic and rapid the change.

4. Your awareness, understanding, and knowledge must be based on the Truth that God, as your ALL-IN-ALL, is right where you are, individualized as you, appearing as you. This means that you do not have to go far to find the Whole Spirit of God. You only have to take your mind off the illusions of "this world" and turn within in contemplative meditation. Very soon you will sense an infinite Knowingness and your emotional nature will tingle with warmth and love. As you continue on the inner journey, the whole vibration of your being changes and you become a transparency for the Light. Now Spirit is "released" to go before you to "make all things new."

5. But sometimes before your world becomes a reflection of this inner Perfection, you shut the door to Spirit and take over the controls once again. So whether you have a total adjustment in your life depends on how long you can keep the door open . . . how long you can allow the Master Consciousness within to work before the little self jumps in and says "my turn" . . . how long before the ego gets in the way.

6. Until there is a full realization and consciousness "locks in" to the Christ Vibration, you must watch your every step. It is much like balancing a jug on your head. When you keep your mind on the Presence within, you are in balance. But if you take your mind off the inner realm of Spirit and place the focus on the world of effect, the slightest stumble in consciousness will topple the jug and you may have a few shattered prayers. The problem is mental laziness. It comes down to a matter of priority.

7. When we look at the priority system of many of the evolved ones, we see that it is quite different from those still operating out of the lower consciousness. Most began just where the majority of people are today, but they began with

dedication and commitment—saying in effect: "I refuse to accept anything but perfect harmony in my life. I will not be sick. I will not be poor. I will not accept inharmonious relationships, and through the Spirit of God I AM, I will shatter these illusions and build a new model of Reality . . . a new world of peace, joy, abundance, wellness, loving relationships, true place success, fulfillment . . . a heaven on earth." And so they established the priority of achieving mastery over this world, and they never lost sight of this goal. Many even made a covenant with God, written, dated and signed on a particular day.

8. Are you ready to take your vows to achieve mastery? This does not mean that you have to withdraw from the world and live like a monk. It means that you begin to really *live* life as the delightful child of God you are in Truth! And your Covenant with the Spirit within does not have to sound like an agreement written by a team of corporate attorneys. Make it simple, but make it meaningful, and I will assure you that if you will keep your part of the agreement, you will marvel at the miracles taking place in your life.

9. Write your covenant according to the dictates of your heart, but here is an example that may help you: "I agree from this moment on to do my very best to live according to the Christ Standard . . . to keep my mind on Him within, to feel love and joy, to think loving thoughts toward all, to speak as the voice of the Master Self, and to act always from a sense of inward direction. To accomplish this, I now release all fears, concerns, resentment, condemnation and unforgiveness. I let go of all negative and fearful images. I surrender all past mistakes and errors in judgment, and I empty out all false pride and ego-centered emotions. Everything in my consciousness that could possibly hold me in bondage I now cast upon the Christ within to be dissolved. I now choose to live under Grace, to be the perfect open channel through which Divine Love, Wisdom and Power flow forth as the Activity of Spirit in my life. And I see and know this Activity to be the perfect harmonizing of all relationships, the perfect adjustment in all situations, the perfect release from all

entanglements, the perfect supply for abundant living, the perfect health of my body, the perfect fulfillment in my life. I now go forth in faith, putting my trust in the Christ within, and living each moment with a heart overflowing with gratitude, love and joy."

10. You keep your part of the agreement by thinking, feeling, speaking and acting in accordance with what you have written. And you practice the Presence throughout each day . . . seeing yourself as God in expression, affirming that the Master Consciousness indwelling is now appearing as your all-sufficiency, seeing your good from your highest vision, and speaking the word that the activity of Spirit is the only Power at work in your life.

11. Remember that you must not only practice the Presence (see the Christ) in yourself, but in all others, too. Understand that there is only one Self in all the universe, one Selfhood—and this Selfhood appears as you, as me, as each individual. Therefore, each and every Soul throughout the universe is a spiritual being—and since my I AM is your I AM, whatever I am saying about you, I am saying about me. If I criticize you, I am criticizing myself. If I see you as poor and weak and unfulfilled, I am seeing myself as poor and weak and unfulfilled. And what I see or say about myself is conditioning my consciousness accordingly. Because of the way the Law works, if I believe that you are suffering from any kind of lack, I am calling for an experience of lack in my life. If I judge by appearances that you are ill, I am setting up the possibility for illness to manifest in my body. Now do you understand why we must not judge others—and why it is absolutely imperative to love thy neighbor as thyself?

Spiritual Activity

Whenever a "need" comes to your mind, identify it consciously and then turn within and recognize that your Higher Self is the answer. Begin to associate everything that you could possibly desire in the outer world with Spirit within as immediate and total fulfillment. This shift in focus from effect to Cause will help to clear the channel for the activity of Spirit. Remember, the Power can work for you only as It works through you.

For example . . .

If you need or desire . . .	Contemplate Spirit appearing as
Money	Lavish abundance, an all-sufficiency of supply with a divine surplus.
Wellness	The restoring Power in every cell and organ . . . the pure and holy Life of the body; energy, vitality, wholeness.
A new loving relationship	The way, means, circumstance and situation for the attraction of the right person, at the right time and right place . . . the drawing together of Souls in an ideal relationship.
To heal a relationship	The energy of Unconditional Love joyously healing and harmonizing the situation.
A job	The perfect opportunity for you to find the greatest fulfillment in providing service to others.

It is also recommended that you write your Covenant with God this very day—and begin working immediately to keep your part of the agreement.

CHAPTER NINE

You Are So Much More Than You Think You Are

All is God, God is Spirit, and Spirit is All. Does this mean that there is no place where the Spirit of God leaves off and man begins? In answering this question, consider that Spirit is Infinite, which means no limitation . . . endless, boundless. We also think of Spirit as Omnipresent, being fully present in all places at all times; Omniscient, possessing total Knowledge, All Wisdom; and Omnipotent, the Almighty, the All Power. Now consider that Spirit is Infinite, Omnipresent, Omniscient, Omnipotent ENERGY . . . energy being defined as the vitality of expression.

2. Imagine now that this All-Knowing, All-Powerful Energy, everywhere present, is pulsating to a particular Divine or Supreme Vibration of Mind. Since it is the nature of this Universal Mind Energy to express Itself, can we not intuit the idea that *as* this Omnipresence begins to conceive of Individual Being— *You*—the vibration of this Omnipresent Energy Field begins to change at the point of conception?

3. Think along these lines for a moment. The Universal Spirit gives "birth" to a new vibration within Itself, and the resulting Consciousness is Spirit being Self-conscious of Itself as Individualized Being— *You*! The word "individualized" means an *indivisible* entity. Therefore, the Universal Presence

cannot separate Itself, or break Itself into parts, in order to become Individual Being. It simply alters Its vibration, and *You* come forth in Consciousness as a particularization of God. Yet the endlessness, the boundlessness, the continuity of Spirit remains the same.

4. Can you understand now that the only "difference" between the Universal Spirit of God and the Individualized Spirit of God *you are* is but a change in vibration? It is a stepping down in frequency to where the Universe says "I AM." This I AM is God . . . this I AM is You . . . Universal *and* Individual Consciousness . . . God knowing Itself as God, God knowing Itself as You, and You knowing Yourself as God.

5. The Ancients taught that this Reality of You, your God-Self, forever remains in the Absolute, and in order to express the cosmic urge of Its Infinite Will, there must be a channel or vehicle for expression. Therefore, your Spirit conceived within Its Mind the Idea of Itself in expression as a living Soul. This has been called "the second creation" where you became a self-conscious entity, created in the Image and Likeness of your Self, the Spirit of God. However, rather than separating Itself, which It could not do, your God-Self followed the original creative Process and changed the vibratory rate within the center of Its Individualized Energy Field. In this "pressed out" state of consciousness, you knew yourself to be a spiritual being, a Son of God, forever living within the Mind of Spirit and filled with the pure awareness of your God-Self.

6. When you took on a body, either by projection or manifestation, the creative energy of your God-Mind was stepped down in vibration for both the non-physical and physical bodies. Now do you see that Spirit, Soul and Body are all the Energy of God in differing degrees of vibration—and that it is impossible for there to be any line of demarcation between God and man? There is no place where God leaves off and man begins. All is God and God is All!

Spiritual Activity

The only "separation" between you and God is the *belief* in separation. To replace that false conviction with Truth, spend time each morning and evening with the following meditative exercise.

Before you begin, understand that there are seven energy centers in the etheric body called chakras, each representing different levels of consciousness. In the treatment, focus your attention in the area of the particular chakra and meditate on the corresponding idea for several minutes—then move up to the next level. This exercise is not related to the esoteric work of awakening the energy centers. It is simply a method used to transmute the energies of lower mind to a higher frequency—into the Christ Vibration where there is no sense of separation.

Say to yourself: **I am much more than I think I am.**

Now focus on the appropriate chakra and think on these thoughts:

Root Chakra—near the reproductive organs: **I am more than a physical body.**

Spleen Chakra—near the navel: **I am more than personality, more than the thoughts of my mind.**

Solar Plexus Chakra—near the solar plexus: **I am more than feelings and emotions.**

Heart Chakra—in the heart area: **I am the unconditional Love of Christ in expression as a Living Soul, and the fullness of Spirit dwells in me.**

Throat Chakra—**I am the Christ Consciousness of Power and Dominion. I am the Creative Master of my world.**

Third Eye Chakra—between the brows: **I am the Christ of God in whom the Father is well pleased. I am Illumined. I see only the Reality of God.**

Crown Chakra—above the top of the head: **I am one with the Universe. I am the Universe. I and my Father are one. All that the Father is, I am. I am the Spirit of the Living God.**

CHAPTER TEN

Working in the Energy of the Absolute

1. In realizing the fulfillment of desires, you must have the consciousness for the thing desired. Without the consciousness, the thing cannot come to you; with the consciousness, it *must* come. As Emmet Fox has written, "The secret of successful living is to build up the mental equivalent that you want; and to get rid of, to expunge, the mental equivalent that you do not want." Remember that a "mental equivalent" is a conviction, a subconscious pattern, a realization, a subjective comprehension of Truth.

2. The first step in building a mental equivalent is to recognize that the Divine Idea corresponding to your desire is already within your Superconsciousness. Think of it this way: The spiritual prototype of everything visible is a part of the energy field that is around and within you. In essence, you already have — *right now* — everything that you could possibly desire, not only for this life, but for all eternity. For example, money is a spiritual idea, as is food, clothing, shelter, transportation, the right work, the perfect body, the ideal mate, and everything else that is manifest as visible form and experience. And all of these spiritual ideas are part of you *now* . . . they are yours *now*, just waiting to express in mind and then in the physical world. Remember that everything is stepped down from the spiritual

to the mental to the physical.

3. In order to build a mental equivalent, there must be a fusion of thought and feeling along the lines of a particular desire-fulfillment. To set the stage for this "fusion" — recall from the previous chapter that your entire being is the energy of God in differing degrees of vibration. There is no place where God leaves off and you begin. All is God. In the individualizing process of Spirit and Soul — and manifestation into form — three distinct vibrations of energy actually came forth — all in unity, all pulsating in an individual force field that comprises your individual being.

4. Think of the first level of energy as the ABSOLUTE, the pure life, pure love and pure intelligence of your God-Self, the Reality of you. This is the realm of Cause.

5. The second level of energy is the ENERGY OF ACTION. This is the karmic phase of mind that works with the Law of Cause and Effect. It is the creative power of the lower planes, and functions as the "laboratory" where mental equivalents are formed. In modern psychology, it is called the subjective or subconscious mind.

6. The third level of energy is the ENERGY OF AWARENESS, or the energy of the relative consciousness, commonly referred to as the objective mind.

7. When the *action energy* — or subconscious mind — receives instructions direct from the *awareness energy* — or objective mind, it will begin to form subjective patterns based on relative conditions, qualifications and precedents. For example, let's say that you are treating for success and prosperity, and you affirm that you are now in your true place with an all-sufficiency of supply. If you affirm primarily out of a relative consciousness, you will be giving the creative power every kind of restriction that may happen to be in your consciousness at the time. In fact, you may be thinking something like this: "I am affirming great success and prosperity in my life, but I really know that there are only certain channels through which my good can come, and I believe that only a certain amount of that good can come forth at this time, and there will probably be a delay because my horoscope says that my houses of career success and

finances are not in the right spots." You can see what kind of mental equivalent you will get with that kind of thinking.

8. When you treat or affirm out of the lower level of relative consciousness, your demonstration will always be dependent on your preconceived ideas. You will do your very best to think of ways that your good can be restricted—or you will consider only one channel through which your good can come—and your reasoning mind may tell you that there is no way the good can come today. But when you tune into your Higher Self and take on the Christ Vibration, you are literally moving out of the relative energy and into the energy of the Absolute. And when you speak the word out of this higher consciousness, your subconscious will comprehend the Truth and will establish the patterns on the basis of *no* limitations, *no* restrictions, *no* time element, and *no* past history. It will operate strictly on the basis of Principle, the Principle of Fulfillment and Abundance that is already your true nature. It will accept the spiritual prototype from the Absolute and duplicate that spiritual idea as a mental equivalent—rather than take your preconceived ideas and build sandcastles that will wash away.

9. When you speak from and as the Absolute, the pattern through which the creative energy of God-Mind radiates is perfect, and as this energy flows through the perfect pattern, it takes on all the attributes of the Ideal and goes forth into the outer world to manifest corresponding circumstances, experiences and form. Spiritual ideas from the realm of Cause are what we are seeking when we pray— "Thy kingdom come. Thy will be done, in earth as it is in heaven." And this is also the meaning of Psalm 127: "Except the Lord build the house, they labor in vain that build it."

Spiritual Activity

Prior to the use of your affirmations today—whether for health, wealth or success—take time to go apart, quiet the mind, let go of the appearances of the outer world, and meditate on the Wholeness and Completeness of your Divine Self. Use this idea as you take the inner journey: **"Closer than breathing is the Presence of God I AM . . . absolute Harmony, perfect**

Love, infinite Wisdom . . . the only Power, the only Cause, and the only Activity of my eternal life."

Contemplate the meaning behind the words and let the spiritual vibration fill your consciousness—then listen as the Voice within speaks to you of your divinity, your holiness.

Continuing in this high state of consciousness, speak the Word for your good, affirming total fulfillment and calling forth into visibility that which you desire.

CHAPTER ELEVEN

Reaching Mountaintop Consciousness

1. What is the most effective way to scale the mountain? Jesus gave us the answer when he talked about the commandment that is the first and greatest of all. He said "Thou shalt love the Lord thy God with all thy heart, and with all thy soul, and with all thy mind, and with all thy strength." Now the Lord *your* God is the Spirit of God within *you* — and you are told to love this Presence with your entire being — with everything you've got. Do you know what it means to love something totally? It means to have constant and continuous adoration for that something . . . to be so filled with devotion, affection, tenderness, warmth, admiration, rapture and love toward that something that your entire consciousness is taken over by it.

2. "That's nice," someone may say. "I may try that someday." Well I certainly hope so, because when you do, everything in your life will change. Understand this: When you contemplate that Presence within, your very Spirit, with great love, that one-pointed love-focus will literally draw the awesome and incredible Power of the universe right into your thinking mind and feeling nature. You take on the Power and

you become the Power and you speak as the Power — and behold — all things are made new! Your thoughts of abundance produce abundance, your feelings of wholeness produce wellness in your body, your vision of success is manifest, your words of love bring forth the relationship you've been seeking.

3. When the Spiritual Masters of the past told us to love the true nature of our being, our Spiritual Reality, only a few people on this planet understood that this was a secret formula for health, wealth and happiness. Only a handful interpreted it from the standpoint of practical everyday living. Only a small percentage recognized it to be the combination to the Storehouse. To love your God-Self is the greatest commandment because within it are the secrets of the universe. And you can track this master teaching all the way back to the beginning of the Mystery Schools. It was taught by all the seers, sages, saints and master souls of the distant past — and one of the most comprehensive teachings is found in Deuteronomy.

4. In Deuteronomy 6:5-9 we are told: "And thou shalt love the Lord thy God with all thine heart, and with all thy soul, and with all thy might. And these words, which I command thee this day, shall be in thine heart: And thou shalt teach them diligently unto thy children, and shalt talk of them when thou sittest in thine house, and when thou walkest by the way, and when thou liest down, and when thou risest up. And thou shalt bind them for a sign upon thine hand, and they shall be as frontlets between thine eyes. And thou shalt write them upon the posts of thy house, and on thy gates."

5. Talk about dedicating you life to the love of Christ within! And when you do, your world will change dramatically. You see, when you direct your attention within and focus that feeling of intense love toward your Higher Self, the entire vibration of your energy field is lifted up to be in tune with the Divine Vibration. And when this happens, you become a transparency for the Activity of God. The radiating Power from within will then dissolve old error patterns, false beliefs and negative appearances, and will move through you

to appear as every needed thing in your life. And the Spirit within is so practical! If you need a better job, one will be attracted to you. If you need more money, it will come in streams of abundance. If you need a healing, the wholeness will be manifest in your body. If you need a new relationship, the right "meeting" will be arranged. You will be shown that there is the perfect supply for every demand.

Spiritual Activity

Let's take the instructions from Deuteronomy and put them into a nine-step program for daily living.

Step No. 1: We are told that these words (loving the Spirit of God within) "shall be in thine heart." This means to impress your subconscious mind with your love of God through daily meditations—by contemplating your inner Being with all the love, adoration and emotion you can feel. So spend time each day in "tender passion" with your God-Self.

Step No. 2: We are told that "thou shalt teach (the love of our God-Self) diligently unto thy children." Your "children" are your thoughts, so you begin now to train your mind— teaching it to relate only to the One Presence and Power within—and to love your Christ Self joyfully and gratefully. Whenever a negative thought enters your mind during the day, gently but firmly say "I choose to control my thoughts by laying aside every weight and turning within to the One I love with all my heart, mind and soul." Then think on that Inner Presence with great love!

Step No. 3: We are told to talk of this love of our Christ Self "when thou sittest in thine house." The "house" is your consciousness, and to "sit" in consciousness refers to prayer or spiritual treatment. Therefore, you begin your spiritual activity by first focusing your love on the Christ within and drawing that reciprocal Love Power into your consciousness.

Step No. 4: We are told to love our Higher Self "when thou walkest by the way." This means that even in moments of idle thinking—when you are simply contemplating the activities of "this world"—you are not to forget the courtship of your True Self. The dominant trend of your thoughts must now be in this direction, regardless of what you are doing.

Step No. 5: We are told to focus that love "when thou liest down." In other words, before you go to sleep at night, again express your deep feelings of love for your beautiful Christ Self. Just say, "I love you so very much. You are so fine, so wonderful, and my love for you fills my entire being to overflowing."

Step No. 6: We are told to express that love "when thou risest up." Train yourself to begin each new day by acknowledging your Higher Self and pouring out all the love you can feel in your heart toward that Master Consciousness within. Say: "Through my love for you I dedicate this day to you. I seek only your will, your word, your way, your work. I let my light so shine this day that I only glorify you."

Step No. 7: We are told that we shall bind this love of our Higher Self "for a sign upon thine hand." Now the "hand" stands for the expressing of God's ideas in the material world—and we are the channels for that expression in our daily activities. Therefore, we are told to bind—or insure—that we make our daily work a symbol or a sign of our love of God. So do everything that is in front of you to the very best of your ability—if for no other reason than for the love of your God-Self.

Step No. 8: We are told that this love for our Inner Self "be as frontlets between thine eyes." That is a direct reference to the creative imagination faculty within all of us, so words and feelings of love for your Christ Self will lift up your vision, expand your consciousness, and enable you to see with new clarity and spiritual understanding.

Step No. 9: And finally, we are told that we shall write these words of love for our Higher Self "upon the posts of thy house, and on thy gates." In other words, keep the love of your God-Self right in the forefront of your consciousness—right within your thinking/objective mind—moment by moment, hour by hour, day by day.

If you will make these nine steps an integral part of your daily living, your life will never be the same. You will be a new creature — "alive with God and upheld by His free Spirit forever!"

CHAPTER TWELVE

Health and Healing

1. All of creation — the infinite universes and all that is visible and invisible — is energy in motion. It is the Thinkingness and Knowingness of God-Mind — Divine Ideas in a state of continuous manifestation. The Spirit of God is pure Cosmic Energy, and this spiritual Substance is individualized as each man, each woman. Therefore, each one of us is an energy field pulsating to a divine vibration. This is our Life Force — the pure Energy of God — and as this Energy lowers its rate of vibration, physical form takes place, manifesting as cells, tissue and organs according to the Perfect Body Idea (the Word). And the Word is made flesh.

2. Ideas such as sickness, disease and old age do not exist in the Mind of God. Therefore, as the pure Energy of God-Mind expresses as the Life Principle and forms the body according to the Perfect Pattern, the visible manifestation must also be perfect. Since we were created out of Perfection, we must be perfect. But how do we explain the appearance of disease and sickness? Go back to the principle that in the Mind of God, Thoughts are creative, and since we are individualizations of God, our thoughts are also creative. We have the freedom to create conditions and experiences in our lives according to the thoughts we think and accept as true.

Thus, we create our own diseases by objectifying fear, hate, worry, or other mental disturbances. But we can also be restored to our normal state of perfection through the Right use of our minds.

3. Any idea that is registered as a conviction in our deeper mind results in a change in our world, beginning with the body. When we begin to consider that the Healing Principle within is the CAUSE of our physical well-being, the negative energy within our individual force field begins to change. In other words, physical perfection is the natural state of our being, and as this Truth is accepted in our thinking and feeling natures, our bodies will change accordingly. So a "healing" is simply a return to our natural state.

4. Based on The Quartus Foundation's research into the subject of health and healing, we believe that an individual can return to his/her natural state of perfection by working with the four "bodies" that comprise individual being: the spiritual, emotional, mental and physical bodies. In the spiritual realm we dedicate ourselves to realizing our True Nature by working from the vantage point that we are NOW spiritual beings — to awaken to the truth of our Divine Image. This is the purpose of meditation, where we dwell upon our Inner Reality, knowing that whatever we contemplate is drawn into our consciousness. This focus on the Christ within, the Spirit of God, will also begin to awaken the subconscious to "remember" the true Image of the Self — the Divine Perfection that we are. So meditation is the foundation for both a restoration and a preventive "medicine" program. Through meditation you will be raising the vibration of your energy field to the divine frequency, thus opening the way for the healing currents to move through every atom of your being.

5. In working with the emotional body, do whatever is necessary to immediately rid yourself of all negative feelings such as unforgiveness, resentment, criticism, fear and jealousy. Even the American Medical Association is talking about the cause and effect relationship between emotions and wellness. In the January 14, 1983 issue of The Journal of the American Medical Association, it was reported that ". . . investigators

found that gum-disease patients had experienced more nega-tive, unsettling life events in the previous year than other people . . . they also demonstrated higher levels of anxiety, depression and emotional disturbances."

6. We have found that the use of spiritual treatments can reverse deep-seated emotional patterns and clear a path for the Inner Power to act. For example, if there is unforgiveness in your heart toward ANYONE (a parent thought for arthri-tis, cancer and heart problems), sit quietly and state firmly and lovingly: "I forgive you totally and completely. I hold no unforgiveness in my heart toward anyone, and if there is any-thing in my consciousness that resembles unforgiveness, I cast it upon the indwelling Christ to be dissolved right now. I for-give everyone and I am free!" Work with such statements, adapting and changing the words for any negative emotion, until you feel a sense of release and there is no longer a nega-tive attachment to the person or experience. You can also use the 10-step manifestation process (from the book *The Mani-festation Process*) to eliminate negative feelings and emo-tions. Choose a Master Thought — a Divine Idea — to replace the negative pattern. Accept it with all your heart and embody it with a sense of HAVE. Then see yourself free of the emotional attachment and express a deep feeling of love . . . speak the word that it is done and surrender your entire being to the Spirit within with great thankfulness, and move out into your world as a fearless, flawless and free Child of the Living God.

7. When we come down to managing our thought proc-esses, we are actually working with the mental body. As Louise L. Hay says in her book *Heal Your Body* — "Stop for a moment and catch your thought. What are you thinking right now? If thoughts shape your life and experiences, would you want this thought to become true for you? If it is a thought of worry or anger or hurt or revenge, how do you think this thought will come back to you? If we want a joyous life, we must think joyous thoughts. If we want a prosperous life, we must think prosperous thoughts. If we want a loving life, we must think loving thoughts. Whatever we send out

mentally or verbally will come back to us in like form. Listen to the words you say. If you hear yourself saying something three times, write it down. It has become a pattern for you. At the end of a week look at the list you have made and you will see how your words fit your experience. Be willing to change your words and thoughts and watch your life change. It's your power and your choice. No one thinks in your mind but you."

8. Remember that the use of creative imagination and visualization techniques also relate to your mental body and greatly influence the physical organism. *See* yourself well! Visualize your wholeness, the natural state of your being. Cancer patients, for example, are benefitting from what is called "positive image therapy." It combines relaxation techniques with teaching the patient to imagine his body's natural cancer-fighting forces — his white blood cells, for instance — and that his cancer is vulnerable to the treatment. In a study at the Washington School of Psychiatry, six patients led by Dr. Robert Kvarnes, had blood samples analyzed before and after the training. The result was that the number of white cells and the amount of a chemical called thymosin in their blood increased. Both changes indicated that their immune systems were stronger.

9. Regarding the physical body, I believe that we must always work from the standpoint of where we are in consciousness—and to not "gamble" by taking action that is beyond our belief system. What I am saying is this: God works through both the metaphysician and the physician. However, healing cannot be complete until the negative patterns in consciousness are corrected. Therefore, medical assistance may offer only temporary relief. Also, a doctor may not be necessary if the individual will combine spiritual work with a good physical health program, i.e. the proper diet, exercise, and good judgment in the maintenance of the body. Nutrition experts can give you valuable information on vitamins and minerals, and excellent books on physical fitness can be found in every bookstore. Rather than advise you personally on these particular "outer" activities, I suggest that

you (1) go within for specific guidance regarding your own situation and what is needed in the manifest world to maintain your body in top physical condition, and (2) follow that guidance to the letter and establish your special health program of foods, supplements, exercise, body cleansing, fasting, natural substitutes for drugs, etc. We each have to find what is RIGHT for us — *individually.*

Spiritual Activity

Let's base our spiritual work on bringing the spiritual, emotional, mental and physical bodies into perfect alignment.

In paragraph 4, we discussed the effects of meditation as a "foundation for both a restoration and preventive 'medicine' program." The form of meditation we are recommending here to realize the true nature of Wholeness is called a "meditative treatment." If you are experiencing a health problem, it means that there is a false belief in your consciousness that is outpicturing itself as dis-ease in your body. There is simply a misconception and a misunderstanding in your mind regarding the natural state of your being. To meet this challenge, you must replace the error with Truth in consciousness, and this can be done most effectively through this type of meditation.

This is the statement that we will work with in our meditative treatment:

The Spirit of God is the Life Force within me, and every cell of my body is filled with the intelligence, love and radiant energy of God-Mind.

God's will for me is perfect health, and God sees me as perfect; therefore wellness is the natural state of my being.

Ideas such as sickness, disease and old age cannot exist in the Mind of God. That Mind is my mind, so I now see myself as God sees me . . . strong, vital, vibrant, perfect. I am now lifted up into the Consciousness of Wholeness. I accept my healing. I am healed now! And it is so.

Now become very still and relaxed — then slowly and with

113

feeling, read the statement again, meditating on each word, contemplating each sentence until the true meaning registers in your consciousness. Remember that words are only symbols . . . it is the idea behind the word that has power. So you meditate on the *idea* until there is an inner understanding and realization. I will lead you through the first meditation, but in subsequent treatments, let your own thoughts replace my words.

Meditation:

The Spirit of God (Contemplate the idea — the meaning-behind the words "The Spirit of God" until you feel something within. Speak the words silently and watch the other thoughts that flow in to expand your thinking.)

is the Life Force within me (Dwell on the meaning and the activity of the Life Force of God operating in and through your body. Feel the dynamics of this incredible power. Sense the renewing, restoring action of Spirit as it eliminates everything unlike itself in your body.)

and every cell of my body is filled with the intelligence, love and radiant energy of God-Mind. ("See" each cell pulsating with Light and Life — and filled with God-Intelligence, God-Love and God-Energy. Each cell is now thinking the Thoughts of God, expressing the Love of God, and vibrating in harmony with the Peace of God. Contemplate this!)

God's will for me is perfect health (Think of God's will as the cosmic urge to express perfection, which is being done in your body right now.)

and God sees me as perfect (This is the Vision of God projecting the Reality of Perfection throughout every cell, organ and tissue of your body. Ponder this!)

therefore wellness is the natural state of my being. (What God sees IS THE REALITY behind the illusion. This Divine Vision, this Holy Seeingness is permeating your entire being. Feel this!)

Ideas such as sickness, disease and old age cannot exist in the Mind of God. (If such ideas do not exist, they cannot be manifest, therefore it is *your* ideas that have been expressed as a negative physical condition. You are now aware of this, and you know that you have the divine authority to replace those error thoughts with Truth Ideas, and you now make the definite decision to do so.)

That Mind is my mind (There is but one Mind — God-Mind. That Mind is in expression as your mind. Your mind, being a part of God-Mind, has the Holy Power of Spirit. And you are now using that Power in cooperating with God. Contemplate God's Mind expressing as your mind, and your mind expressing God's Ideas of Perfection.)

so I now see myself as God sees me . . . strong, vital, vibrant, perfect. (Lift up your vision and see as God sees. See Wholeness. See Wellness. See Divine Order. See Perfection. See God *as* your body!)

I am now lifted up into the Consciousness of Wholeness. (Feel the pure vibration of Love, Life and Light as you rise into the very Presence of Spirit. Meditate on the spiritual energy that now surrounds you, engulfs you, and flows in and through you. Let go and give yourself to the magnificent Healing Currents.)

I accept my healing. I am healed now! And it is so. (When you accept your healing, you have taken the final step. Where there was darkness, there is now Light. Where there was error, there is now Truth. Where there was imperfection, there is now Perfection. You are healed! Acknowledge now that it is so!)

Remain in the consciousness of Spirit for a few more minutes — in communion with your God-Self. In this spiritual vibration you will be highly successful in dealing with your emotional body. Forgiving others will be easy, and old hurts, resentments and other negative feelings can quickly be cast upon the Christ within to be dissolved. For this particular activity, make a list of everyone who could possibly need your forgiveness, then speak their name aloud and say: "I forgive you. I choose to do this now, and I hold nothing back. I forgive you totally and completely!" Next, take an imaginary box and in your mind, fill it with every hurt, resentment, condemnation, depressed feeling, anger thought, and any other negative patterns you find in consciousness. Now take the box and see yourself lovingly placing it upon the Holy Fire of Spirit within where it is totally consumed.

To properly manage your thought processes, refer back to paragraph No. 7 in this chapter and begin to listen to the words you say throughout the day. What habit patterns are you forming? Start exercising control over the thoughts you think and the words you speak. *Practice* thinking joyful, loving, prosperous, and harmonious thoughts. Train yourself to think and speak only according to the Christ standard, and use your power of creative imagination to see yourself as whole, well and perfect.

In working with the physical body, ask yourself: "What do I intuitively feel that I must do in the manifest world to maintain my body in top physical condition?" You may be told to relax more, to eat only those foods that you *know* are appropriate for you, and to exercise regularly. Whatever the answer may be, just be sure to follow your inner guidance in establishing a "health program" that is right for your individual consciousness.

Work daily to keep your four bodies in holy agreement, and sickness will be a thing of the past for you.

CHAPTER THIRTEEN

The Principle of Abundance

1. As you gain the understanding of the Principle of Abundance, and realize (know) the Truth therein, you will be free of all lack, limitation and imperfection, beginning with your body and continuing out to encompass all conditions, situations, circumstances and experiences of your life and affairs. Reason: Your outer world will be a reflection, or an outpicturing, of Truth, rather than the false beliefs of the carnal mind.

2. Do you have an aversion to wealth? Do you object to being rich? Does the word "abundance" bother you? If you say "yes" — then you do not believe in God, because God is omnipresent Wealth, the infinite Riches of the universe, the lavish Abundance of creation. And if you deny unlimited prosperity, you are denying yourself, because YOU are the image of omnipresent wealth, the expression of the infinite riches of the universe, an individualization of lavish abundance. You are as rich *right now* as any individual who ever walked on this planet. The cattle on a thousand hills are yours, the gold and silver are yours, and an abundance of money is yours now!

3. If you do not believe that God loves prosperity, just read your Bible.

• Beloved, I wish above all things that thou mayest prosper.

- Prove me now herewith, said the Lord of hosts, if I will not open you the windows of heaven and pour you out a blessing, that there shall not be room enough to receive it.
- The blessing of the Lord, it maketh rich, and He addeth no sorrow with it.
- They shall prosper that love Thee. Peace be within thy walls, and prosperity within thy palaces.
- God is able to make all grace abound toward you; that ye, always having all-sufficiency in all things may abound to every good work.
- . . . thou shalt remember the Lord thy God, for it is he that giveth thee power to get wealth . . .
- The Lord shall open unto thee his good treasure, the heaven to give the rain unto thy land in his season, and to bless all the work of thine hand; and thou shalt lend unto many nations, and thou shalt not borrow.
- Let the Lord be magnified which hath pleasure in the prosperity of his servant.
- Therefore I say unto you, All things whatsoever ye pray and ask for, believe that ye have received them and ye shall have them.

4. Your Lord is the Spirit of God, the Christ, within you. You magnify this Master Self that you are in truth by realizing that you and God are one. All that this one Presence and Power of the Universe is, you are—and all that this Infinite Mind has, is yours. Above you, around you, in and through you, is *You* . . .the Reality of You, an omnipotent Force Field embodying all Love, all Wisdom, all Life, all Substance, all ALL. This Allness that is individualized *as* you is the same Mind, the same Spirit, the same Presence who spoke to Moses from the burning bush, the One who spoke through Jesus.

5. This Spirit within you is forever thinking thoughts of Abundance, which is its true nature. Since thoughts of lack or limitation can never be registered or entertained in this Infinite Mind, then the Principle or Law of Supply must be one of total and continuous All-Sufficiency. Your Self thinks, sees and knows only abundance, and the creative energy of this Mind-of-Abundance is eternally flowing, radiating,

expressing, seeking to appear as abundance on the physical plane.

6. This radiating creative Mind Energy is substance. As this Divine Thought Energy flows through your consciousness and out into the phenomenal world to appear as prosperous experiences and conditions, its "plastic" quality allows it to be impressed by the tone and shape of your dominant beliefs. Therefore, what you see, hear, taste, touch and smell are *your* beliefs objectified. The form and the experience are but effects—appearances—and we are told to not judge by appearances. To "judge" something means to believe it, to assume that it is true, to conclude that it is factual. But we are told not to do this. Why? Because what appears as an effect has no value in itself. The only attributes that an effect has are the ones that *you* give it.

7. Money is an effect. When you concentrate on the effect, you are forgetting the cause, and when you forget the cause, the effect begins to diminish. When you focus your attention on *getting* money, you are actually shutting off your supply. You must begin this very moment to cease believing that money is your substance, your supply, your support, your security, or your safety. Money is not—but God is! When you understand and realize this Truth, the supply flows uninterrupted into perfect and abundant manifestation. You must look to God alone as THE Source, and take your mind completely off the outer effect.

8. If you look to your job, your employer, your spouse or your investments as the source of your supply, you are cutting off the real Source. In fact, if you look to any human person, place or condition for your supply, you are shutting down the flow. If you give power to any mortal as even being the channel for your supply, you are limiting your good.

9. You must think of money and any other material desire or possession simply as an outer symbol of the inner supply. And the only Reality of that symbol is the substance which underlies the outward manifestation. Money is the symbol of an Idea in Divine Mind (as is every other good thing). The Idea is an all-sufficiency of supply to meet every need with a

divine surplus in your individual life. As the Divine Idea comes out into manifestation, it appears as the symbol: money. But the money is not the supply. Rather, it is your consciousness of God *as* your abundance that constitutes your supply. When you try to collect, acquire and possess the symbol (focusing on the symbol and not the supply within), the outlet for the manifestation closes.

10. Do you want more money, more prosperity in your life? Then shift from a consciousness of effects (materiality) to a consciousness of cause (spirituality). When you give power to an effect, you are giving it *your* power. You are actually giving the effect power over you. Does money have power? If you say "yes" — then you have given it *your* power and you have become the servant. You have reversed the roles.

11. The Inner Presence — the You of you — is truly the money-maker. Your thinking, reasoning mind is not. Your *only* Source is the God Presence within you. If your mind is on the Source, the Cause, the supply flows freely. If your mind is on the effect, you block the flow. The more *impersonal* you become regarding *where* your money seems to originate (job, salary, commissions, investments, spouse, etc.), the more *personal* you can become in your relationship with the true Source of your money, and the closer the relationship to your God-Self, the greater the abundance in your life.

12. Turn within and watch the Inner Presence work. The activity of your Infinite Mind sees and knows only abundance — and in this sea of Knowingness is a spiritual Idea corresponding to every single form, event, circumstance, condition or experience that you could possibly desire. The creative energy (substance) of these Divine Ideas is forever flowing into perfect manifestation. But remember, if you constantly look to the effect, the visible form, you will create a mutation, a less-than-perfect manifestation. By keeping your focus on Spirit, however, you will keep the channel open for the externalization of Spirit according to the Divine Idea.

13. The time must come when you will satisfy a need for money by steadfastly depending on the Master Self within —

and not on anything in the outer world of form. Until you do this, you will continue to experience the uncertainties of supply for the rest of your life. Every soul *must* learn this lesson, and until it does, it will be given opportunity after opportunity in the form of apparent lack and limitation. You may be experiencing such a challenge right at this moment. Realize that this is the opportunity you have been waiting for to demonstrate the Truth of your birthright. Know that this entire experience is but an illusion, an outpicturing of your beliefs, an effect of your consciousness. But you are going to stop giving any power to the illusion, to the effect. You are going to cease feeding it with negative energy. You are going to withdraw your energy from the outer scene and let it die, let it fade back into the nothingness from which it came.

14. Take your stand this day as a spiritual being, and renounce all claims to humanhood and mortality. Care not what is going on in your world, regardless of your fears about your creditors, your security, your protection, your future. Turn away from the effects, wave good-bye to external false-belief pictures, and return to the Father's House where you have belonged ever since you left under the spell of materiality. Take your mind off money and material possessions (the effects) and focus and concentrate only on the lavish abundance of divine substance that is forever flowing from that Master Consciousness within you. Take your stand and prove God now!

15. Stop adding up your bills, stop counting the money you have or need, and stop looking for your supply from any mortal person, place or situation. The whole Universe is standing on tip-toe watching you — praying that you will let go of the negative appearances of the world of illusion and claim your divine heritage. Now is the time—today is the day. Pass this test and you will never have to go through it again. But if you yield to mortal pressure and carnal mind temptation to get temporary financial relief from the world of effects, you will have to go back to the classroom and learn the lesson all over again.

16. Say to yourself with great feeling: "This day (speak the

actual date) I cease believing in visible money as my supply
and my support, and I view the world of effect as it truly is .
. . simply an outpicturing of my former beliefs. I believed in
the power of money, therefore I surrendered my God-given
power and authority to an objectified belief. I believed in the
possibility of lack, thus causing a separation in consciousness
from the Source of my supply. I believed in mortal man and
carnal conditions, and through this faith, gave man and conditions power over me. I believed in the mortal illusion created by the collective consciousness of error thoughts, and in
doing so, I have limited the Unlimited. No more! This day I
renounce my so-called humanhood and claim my divine
inheritance as a Be-ing of God. This day I acknowledge God
and only God as my substance, my supply and my support."

17. Now impress these statements of Principle on your
mind:

- **God is lavish, unfailing Abundance, the rich omnipresent
substance of the Universe. This all-providing Source of infinite prosperity is individualized *as* me — as the Reality of
me.**

- **I lift up my mind and heart to be aware, to understand,
and to know that the Divine Presence I AM is the Source
and Substance of all my good.**

- **I am conscious of the Inner Presence *as* my lavish Abundance. I am conscious of the constant activity of this Mind
of infinite Prosperity. Therefore, my consciousness is filled
with the Light of Truth.**

- **Through my consciousness of my God-Self, the Christ
within, as my Source, I draw into my mind and feeling nature the very substance of Spirit. This substance is my supply, thus my consciousness of the Presence of God within
me *IS* my supply.**

- **Money is not my supply. No person, place or condition is
my supply. My awareness, understanding and knowledge of
the all-providing activity of the Divine Mind within me is
my supply. My consciousness of this Truth is unlimited,
therefore, my supply is unlimited.**

- **My inner supply instantly and constantly takes on form and** experience according to my needs and desires, and as the Principle of Supply in action, it is impossible for me to have any needs or unfulfilled desires.

- The Divine Consciousness that I am is forever expressing its true nature of Abundance. This is its responsibility — not mine. My only responsibility is to be aware of this Truth. Therefore, I am totally confident in letting go and letting God appear as the abundant all-sufficiency in my life and affairs.

- My consciousness of the Spirit within me *as* my unlimited Source is the Divine Power to restore the years the locusts have eaten, to make all things new, to lift me up to the High Road of abundant prosperity. This awareness, understanding and knowledge *of* Spirit appears *as* every visible form and experience that I could possibly desire.

- When I am aware of the God-Self within me *as* my total fulfillment, I am totally fulfilled. I am now aware of this Truth. I have found the secret of life, and I relax in the knowledge that the Activity of Divine Abundance is eternally operating in my life. I simply have to be aware of the flow, the radiation, of that Creative Energy, which is continuously, easily and effortlessly pouring forth from my Divine Consciousness. I am now aware. I am now in the flow.

- I keep my mind and thoughts of "this world" and I place my entire focus on God within as the only Cause of my prosperity. I acknowledge the Inner Presence as the only activity in my financial affairs, as the substance of all things visible. I place my faith in the Principle of Abundance in action within me.

Spiritual Activity

Here is a program for realizing abundant prosperity in your life and affairs. It takes 40 days for consciousness to realize, or develop a subjective comprehension, of a truth. A break during the 40 day period releases the energy being built up around the idea. Therefore, there must be a definite commitment to faithfully follow this program each and every day for

40 days — and if you miss even one day, to start over again and continue until you can go the full period with perfect continuity. Here is the course of action:

- Establish a specific date to start your program, such as the beginning of a particular week. Count out 40 days on your calendar and mark the completion date.
- On the first day of the program, write the statement shown in paragraph 16 in your Spiritual Journal.
- There are 10 statements of Principle. Read *one* statement each day. This means that you will go through the entire list *four times* during the 40-day period.
- After reading the daily statement — either upon arising or before going to bed in the evening — meditate on it for at least 15 minutes, focusing on each idea in the statement with great thoughtfulness and feeling — letting the ideas fill your consciousness.
- Following each meditation period, write down in your Journal the thoughts that come to you. *Be sure to do this daily!*
- If you are working in a weekly Master Mind or study group with other Quartus members, exchange the thoughts written in your Journals and discuss them in the group for greater illumination and understanding.
- Since you have *already* received an all-sufficiency of supply (all that Infinite Mind has is yours now), you can prove this Truth to your deeper mind by sharing your supply on a regular basis. Giving is an esoteric science that never fails to produce results if it is done with love and joy, because the Law will shower you with a pressed down and multiplied return. But if you tithe (and I really prefer the word "sharing'" to tithing) as a mechanical and calculated method to please God, unload guilt, meet a sense of obligation and play a bartering game with the Law, no one benefits — not even the receiver. Give with love, joy and a sense of fun and the windows of heaven will be thrown open with a blast!

CHAPTER FOURTEEN

Your Consciousness is Your Faith

1. The entire Universe is nothing but pure God Energy . . . vibrating, thinking, knowing . . . omnipresent, omniscient, omnipotent. And within this Divine Radiance is the attribute of Absolute Faith . . . total belief, total confidence, total trust, total certainty, total conviction in Itself.

2. As this Infinite Presence and Power of Absolute Fidelity expresses *as* you, as It brings the whole thrust of the Universe into individualization, It also brings with It the incredible Energy of Faith to serve as one of your Divine Powers. This Power is the very foundation of your Soul—the energy of your consciousness.

3. Faith *is* your consciousness. You think, feel, speak and act according to your consciousness, according to your faith. Faith is the substance (the creative energy) of things hoped for, the evidence of things unseen, therefore, your consciousness is that which stands under and supports (the substance of) that which you are experiencing in your world. Your consciousness is the *present* evidence of what you will experience in your life as your thoughts and emotions are externalized.

4. If your consciousness is filled with fear and anxiety, that is where your faith is. You are putting your faith in the possibility and probability of misfortune, lack and limitation. Your consciousness, which is your faith, your substance, must by law act upon itself. Thus, you, as Creative Energy, will create in your world exactly those conditions you fear (have faith in).

5. Everything in your life is vibrating according to a certain pitch and frequency. Your body, home, car, clothes, job, relationships, and money are all energy in motion-all vibrating in absolute exactness to the vibration of your consciousness. Your faith vibration attracts that which you have and experience in this world because *like must attract like!* Where is your faith? Look around you. If your faith is in an all-sufficiency, then so it is in your life. If your faith is pulsating to a "just getting by" frequency, then so it is in your world. If your faith is on the dial-set of insufficiency, then there will never be enough to meet your needs. Your world simply reflects your faith.

6. The faculty of *Divine* Faith (Faith Energy pulsating according to its divine vibration) may represent only a tiny particle of light within your lower mind at the present time. But Jesus said that if your faith was no larger than a grain of mustard seed (or just a faint glow of light), you could level a mountain. Now Jesus didn't kid around. When he made a statement, you could bet your life on it. But then we ask: "What about that mountain of debt, this peak of despair in my relationships, that volcano that has erupted in my body, those rising fears regarding my career?" Could the answer be that you have not recognized this Power Center as being an integral part of your individualized cosmic system? Could it simply be that you are not *aware* that you have this inexhaustible Power right at your disposal?

7. Perhaps you have not been aware of the Energy of Faith, yet you have been using it. You may not have known that you had this Power, but all the time it's been the basic ingredient in all your creative activities—and many of those creations have been of a negative nature. You have created illness

through the energy of faith—faith in drafts, weather, heredity, germs, viruses, old age, decay and disease. And through your power of faith you have created lack and limitation—faith in the economy, in money as your supply, in the intentions of others, in your own scheming and manipulating. You are not alone. We have all created a world of illusion with our faith. Everytime we say that we are afraid of something, we are putting our faith in that something. What we feared most comes upon us because our faith brings it to us. If we believe that anything negative can happen in our lives, then we are vulnerable to it. If we believe in the possibility of accidents, disease, failure, or suffering of any kind, we are lowering the rate of our faith vibration and sending out negative energy to attract misfortune into our lives.

8. When we become aware of our Power of Faith, the energy from that Center begins to work for us according to the Divine Standard, the original High Vibration. Think of it this way: If you *do not know* that you have a certain attribute, a specific power, then your activities and decisions are based on a power outside of you. Through this belief in an outside power, you are actually transferring (giving up) a God-given power that you didn't even know you had—and you are giving it to the so-called "outside forces." These forces then become the master and you become the servant. However, when you begin to recognize that you have an incredible Power Center within your consciousness—one that represents the very Power of God—then you are calling forth the pure form of that energy to be used in transforming your life.

Spiritual Activity

Let's do something about that faith energy that you have been using in negative ways. Say to yourself, aloud and then silently with great feeling: **I am the Power of Faith!** Let that idea roll around in your mind for a few minutes and seep down into your emotional nature. Feel its vibration as your entire energy field begins to strengthen, firm up, become substantial (filled with substance).

Now slowly contemplate these statements:

I believe there is NOTHING God cannot do. I believe there is nothing God cannot do THROUGH me. I believe there is nothing God cannot do AS me. God is my Self, therefore there is nothing that I cannot do. Nothing is too good for God, and nothing is too good for me!

The pure energy of Faith is now flooding your consciousness, transmuting the negative frequencies and restoring your soul to the spiritual vibration. And with this kind of consciousness, you can subdue kingdoms, bring forth righteousness, stop the mouths of lions, quench the violence of fire, and escape the edge of the sword. (See Heb. II, 11:33-34).

The Faith Power is located in the upper part of your energy field corresponding to the center of your physical brain. Accordingly, you use your thinking mind to first contact this Power. See it with your inner vision as being half-way between your eyes and the back of your head, as a small circle of light. As you bring the light into your awareness and begin to focus on it, notice what happens. It begins to intensify, first radiating upward to fill the head and continuing to move up until the entire upper part of your energy field is bathed in brilliant light. Then the light begins to project outward to each side of your energy field, reaching to the outer limits. Now it moves downward until it fills your entire field of consciousness with pulsating energy. Practice this exercise, knowing that as the light radiates from its center position, you are watching pure God Energy move through your mental, emotional and physical systems . . . purifying, transmuting and saturating your being with the High Vibration of Faith.

Following this exercise, bring your point of contact with the Faith Energy to your heart center by feeling the very core of this Power right in your heart. Feel the vibration, and speak these words aloud and then silently:

I love the Faith I AM with all my heart, and I now draw

forth the omnipotence of this incredible Energy and command it to fill my feeling nature with its Power. Come forth, my Faith! Saturate my emotions with total trust, total certainty, total conviction in myself as a Being of God. I AM Substance. I AM Creative Energy. I AM God being me now. Through the pure Energy of Faith, I feel this Truth!

Now bring your point of contact with the Faith Energy up to your throat center, and feel that new strength in your throat. Speak these words aloud and then silently with great feeling:

Through the Power of Faith, I speak the word and it shall not return unto me void. I AM omnipotence made manifest on earth. I AM unlimited. I have the Power, and I use the Power rightly and wisely and lovingly in the name of Almighty God.

Let your point of contact with the Faith Energy now be between your brows, and speak these words aloud and then silently with great concentration:

I AM a master mind, created in the image of God, and I dedicate my mind to the service of God and to all God-Kind everywhere. My mind is power-full, filled-full with the Energy of Faith, and there is nothing that I cannot do.

Move your vision up to just above the top of your head, and speak these words aloud and then silently with great reverence:

I place my Faith in the Reality I AM, the very Christ of God. I now let my world reflect the Divine Activity of love, life and abundance, for my Faith has made me whole.

Work with the Power of Faith daily, knowing that this awesome force will penetrate into the depths of consciousness and burn away all error thoughts and negative beliefs. It is the rock upon which you shall build the foundation for a full, glorious and lasting spiritual consciousness. It is indeed the key to mastery!

CHAPTER FIFTEEN

Spiritual Strength and Wisdom

1. As the Power Center of Faith opens (awakens), its energy begins to radiate and interact with three other Centers with which it has close affinity — Strength, Wisdom and Love. We will discuss Strength and Wisdom now and follow with Love in the next chapter.

2. The Strength we are referring to here is defined as spiritual firmness and mental toughness . . . soundness, boldness, steadfastness in consciousness. It means a solidifying of your awareness, understanding and knowledge of God by eliminating all sense of spiritual weakness and revealing the majesty and magnificence of your True Nature. As the Center of Strength (positioned in your energy field near the physical location of the lower back) is awakened, it generates a feeling of poise, confidence, and great stability.

3. Strength is a "brother" of Faith — and if either one of these Power Centers is lowered in vibration, the other increases its tempo to offset the loss and bring the combined energies back into balance. But this action does not take place unless you are dedicated in your efforts to awaken to your Divine Identity. The development of a spiritual consciousness and the awakening of the Power Centers go hand

in hand. You cannot have one without the other.

4. When you begin to work with your Faith faculty, as discussed in the previous chapter, it automatically interacts with Strength, and you immediately feel that new Power vibration in consciousness. Faith and Strength are now in support of one another, and if a negative condition comes to your attention and your faith drops momentarily, the Power of Strength will rise up and say, "What is that to you? Do not place your faith in appearances!" And your ego counters by saying, "But there's not enough money to pay the bills." And your firm position in Spirit replies: "What do *you* know? I am strong in the Lord I AM, and I will not tolerate such an illusion. I am the Abundance of the Universe individualized, and I choose now to express an all-sufficiency to meet every need with plenty to spare and share."

5. Later the ego says, "I don't feel well . . . I think I'm going to be sick." And your mental toughness replies, "That's a lie! The Spirit of the Living God I AM is my life, and it is impossible for God's Life to be less than perfect. Therefore, I am whole, complete, and wonderfully WELL!" Still later your little me says, "But . . ." and you cut it off instantly — "No buts about it. I take my stand in the Omnipotent Christ I AM, and I permit no false beliefs or negative emotions to enter my consciousness. I refuse to play your silly game any longer!"

6. As Faith and Strength are developed, the Wisdom faculty will open beautifully and you will be lifted up above the level of so-called "common sense." Common sense is fine for the third-dimensional man or woman, but if you continue to work on that level, particularly out of a common sense of fear, you will never achieve mastery. For example, let's say that you must make a decision regarding your job or career, and common sense tells you to stay where you are because of the security. But what does Spirit have to say in the matter? You take the question within and ask for spiritual light and understanding, and your intuition may tell you to be bold and step out in faith — that the new career opportunity is truly the stepping stone to your True Place. A friend of mine

had a "gut feeling" to leave his secure and well-paying job as a CPA and start a new and totally unrelated business. He followed the inner leading — taking a substantial reduction in income — but within a few years he was a multi-millionaire. What he did was the perfect example of uncommon sense!

7. As your Wisdom Center becomes more vibrant, you will know what to do without going through a long and "logical" reasoning process. You will *feel* that a certain action is right — and you will move ahead without hesitation. The difference in good judgment and common sense? One is based on Spirit as *Cause*, while the other tends to draw support from the outer world of *effect*. If you are working with the Powers of Faith and Strength, you will not be fearful and overly-cautious in taking actions — and at the same time, you will not be compulsive or inconsiderate. Your actions will be guided by intuition and the inspiration of Spirit. (The Wisdom faculty is located adjacent to the solar plexus in an individual's energy field.)

8. Remember that it is not your ego who is all wise. As Socrates wrote: "The Delphic oracle said I was the wisest of all the Greeks. It is because that I alone, of all the Greeks, know that I know nothing." And this is true of you, but when your Wisdom faculty is awakened, your consciousness becomes the channel through which the Wisdom, Understanding and Knowledge of the Omniscient Christ Mind flows. And this is what it means to be an Illumined One.

Spiritual Activity
Spiritual treatment for strength:

I am the Power of Strength, I am Power-full. I am strong in the mightiness of Spirit and I am undaunted! My mind is firmly one-pointed in seeing only the good. My heart is fearless and knows only the emotion of victory. Nothing can touch me but the direct action of God and God is my Omnipotent Self. I can do all things through the Strength of the Christ I AM. I AM STRENGTH!

Spiritual treatment for wisdom:

I am the Power of Wisdom, and I call on this Power now to fill my heart and mind with the Light of perfect judgment and intuition. Through Christ in me, the very Spirit of God I AM, my actions are right and perfect. I know what to do at all times and in every situation. And I always do the right thing because it is the right thing to do! I KNOW! I FEEL! And what I know and what I feel are Spiritual Knowledge and Inspiration guiding me every step of the way. God cannot make mistakes and neither can I when I am consciously aware of the Presence within. I am now aware of that Presence, and I am filled and thrilled with the illumination of Spirit. I AM WISDOM!

CHAPTER SIXTEEN

The Power of Love

As we sit here in this imaginary circle, permit me to play back to certain ones of you what you have told me. Perhaps you did not speak these exact words, but to paraphrase Emerson, what you are in consciousness speaks louder than words. This is not judgment, for anything we can see in another, we have seen in ourselves.

1. You think that you can sit around with your nose out of joint and have others run around catering to you. You wear your feelings on your sleeves, and if everyone's actions do not fulfill your rigid expectations, you feel rejected. And you say that you are a practicing metaphysician!

2. And you with the short fuse. You call yourself a Truth student—which is synonomous with practicing the art of loving—but the only art you're developing is how to lose patience, get angry and throw a good tantrum. And you wonder why your life is not whole and complete.

3. And you, the great hugger. You really know how to display affection in public, but the bitterness expressed behind your closed door is enough to keep your blood pressure high and maintain that pain in the neck in perfect order. Can you not love when you are alone?

4. And you, the "evolved" one. You say that you *know*

Truth—and yet there is that deep-deep resentment toward people of the past and present. Evolvement comes through love, and not the other way around.

5. And you with the sharp tongue. How you love to criticize your minister (not to his face of course), but when asked what *your* ministry in life is, you say you can't get on with it because no one understands you. That's because you're not using the universal language of unconditional love.

6. And you, the fence-straddler. The shift of your consciousness remains in neutral, and though you race the engine of your mind and emotions constantly, you go nowhere. And the reason is because you haven't begun to express your love nature, which is the go-power of the universe.

7. I could go on around the circle, but I think you get the point. If your life is not overflowing with abundance, wellness and fulfillment, you are out of tune with the Love Vibration within you. If you want more out of life, you are going to have to give more to life. When you give love, you receive the Kingdom.

8. Love is what created the universe, and Love is what the universe was created out of. Therefore, Love is Mind and also the thoughts of Mind. Love is the thrust of all creation. "And God said . . ." And the Word was Love . . . and the Power of the Word was Love . . . and the manifestation of the Power was Love. All *is* Love!

9. The Infinite All is the pure essence of Love. This Infinite Love thinks. Its Consciousness is perfect Love. As It contemplates Itself, It does so with Love. And what It sees, It loves. This Father-Mother Mind conceived the perfect Image of Love, which became the first Principle, the ever-living male and female Principle, the I AM THAT I AM, the Love-Self-Reality of each one of us. And this Spirit of Love expressed as you, as me, as every living thing, as all that is.

10. In the beginning you knew only Love. And your creations of materiality were born out of love, for you were a co-creator with God, bringing forth into manifestation only the Divine Ideas of Love. But even a Love-Child has free will, and

you chose to create forms and experiences without the counsel of the Father-Mother Love within. And once you began to identify yourself with your creations, you sealed off the Love Vibration with a material consciousness. Yet you continued to be, and will always be, a spiritual being of Love.

11. Some men and women have rediscovered (awakened to) their true nature of Love, and have opened the inner door to receive once again the Energy of Love, letting it fill their consciousness and eliminate the error patterns of the past as light dissolves darkness. We call these people Superbeings.

12. Are you one? If you are, you know that you are Spirit, that Spirit is Love, and that Love is the activity of Spirit. You know that the activity of Spirit is Its Self-expression, and since you are that Self-expression, you are pure Love. And you know that the Love that created you forever sustains you. Oh how your God-Self loves you! You *know* this, and you are eternally conscious of this joyful Truth!

13. Knowing that the Spirit of God within you—*your* Spirit—loves you with all of Its Being, and that you would not exist without the full focus of this Love appearing *as* you—you relax and let the burdens fall from your shoulders. Say to yourself:

Since the only Presence and Power of the Universe loves me and sustains me, what on earth could I possibly fear? Nothing. No-thing. Love heals. Love prospers. Love protects. Love guards. Love guides. Love restores. Love creates. Love makes all things new. So I let Love go before me now to straighten out every crooked place in my life. I place my faith in God's love for me and I am free, as I was created to be!

14. Your God-Self will restore your life and transform your world into a Garden of peace, joy, beauty, abundance and fulfillment. But remember, you are a co-creator with God—not just an empty projector through which images are thrown on the screen of your world. You have a role to play, too, and that role is to be a conscious participant as a radiating center

of Divine Love.

15. Your "center"—which is another word for your energy field—includes thoughts, feelings, words, and deeds. Therefore, to be a co-creator with the Spirit of Love, you must think love, feel love, speak love, and act with love. Your first thought of love should be to respond to the love that your God-Self is eternally pouring out upon you. Since this Presence within you loves you with all of Its Divine Consciousness, should you not reciprocate by loving this Reality of you with all your mind, your heart, your soul, your strength? Can you not express gratitude for that love by returning the love in full measure? When you do, the Connection is restored and the middle wall of partition is blown away.

16. Turn within and say:

Thank you for loving me. Regardless of what I have done in the past, I know that your love for me has never diminished. Even when I have ignored you, or blamed you, or took action contrary to your counsel, you continued to love me with all of your Being. I love your Love! And I love You! My heart runneth over with love for You, my Friend, my Guide, my Wonderful One, my Counsellor, my mighty God, my everlasting Father, my Prince of Peace, my very Christ Self! Oh, I love You with all my mind, with all my heart, with all my soul, with all my strength. Love is what I have received and Love is what I give, and I am now whole and complete.

17. Since this God Presence within you *is* you—the Higher Self of you—and since this Self is forever expressing as the allness of you, including your Soul and body, can you now begin to love *all* of you from center to circumference? There is no place where God leaves off and you begin, so all is God and all is you! KNOW THYSELF! To know yourself is to love yourself—all the way through. And do not think that you are not worthy, because your worthiness is God's Worthiness! LOVE THYSELF!

18. Think beautiful thoughts about yourself:

I am a delightful Child of God. My Spirit is God being me in the absolute. My Soul is God being me in expression. My body is God being me in physical form. I am God being me! Knowing that everything I have ever done, ever said, ever thought, ever felt was simply my consciousness in action, I understand that I could not have expressed any differently. I was acting out of my consciousness, therefore, I dismiss all thoughts of right and wrong . . . that was simply where I was at the time . . . and I know now that I am an evolving Soul returning Home to the Light of Love. So I no longer condemn ME . . . I no longer hold any unforgiveness toward MYSELF. I rise above all feelings of guilt, and I am free to love myself as never before. I love this person I am with all my mind. I love this INDIVIDUAL BEING that I am with all my soul. My love for ME, MYSELF, the I that I AM, knows no bounds. I AM LOVE. I AM LOVE. I AM LOVE.

19. You are told to love your neighbor as yourself, and your "neighbor" means every other soul on this planet and beyond, and all forms of life throughout the universe. So now turn your attention to the world and begin to radiate the Love Activated Spiritual Energy Rays. That's your Laser beam, and when you direct the beam from the Love Energy Center of your heart, it goes before you to transform every negative situation your neighbor may be experiencing into a splendid positive. This is Power-Love, rather than cuddly fuzzy love, and there is nothing it cannot do. When you direct it toward anyone or anything, it literally changes the energy field in and around the person, place or thing. This is God in action, peeling away the illusion and revealing the Reality.

20. Say to yourself:

I will do my part to love my neighbor without exception. As I scan my consciousness, my mind picks up certain individuals of the past and present who evoke less-than-desirable emotions in me. I now transmute that negative

energy by forgiving them and speaking words of unconditional love. (Speak the name aloud) . . . I love you! I love
you unconditionally! I love you for Who and What you
are, with no strings attached. I am love. You are love. We
are one in love, and we are healed through love. I now
bring into my consciousness my home and family, my place
of work, my city, state and country, my world — and I send
forth the Love Activated Spiritual Energy Rays to heal and
harmonize every negative condition on this planet. I feel
the love pouring forth from my heart center, and I know
that this Love Power will accomplish that for which it is
sent. I AM LOVE IN ACTION!

Spiritual Activity

Spend time daily with the affirmative prayers in para-
graphs 13 through 20 — then continue to *be* Love in action!
Be God in action! Do not get caught in the mesmerism of lis-
tening and watching others fight, fume and spew negative
energy. Begin to pour love into the situation from your heart
chakra, radiating it with intensity, and joyfully watch as the
individuals are touched by the harmonizing rays. If injustice
comes within the range of your consciousness, send forth the
spiritual Energy of Love and see Right Action taking place.
Use the Love Power in your home, your office, in the grocery
store, the restaurants, the hospitals, the courtrooms, on the
freeways — and notice how the environment changes. Stop
being a spectator! Use your Lasers in the service of Godkind
to reveal order, harmony and peace in this world.

Practice the use of Love-Power daily and prove to yourself
that you do indeed have a Divine "Zapper" at your disposal.
If an insect bites you, focus the love energy at the point of
contact and feel the instant relief from the sting. If you meet
someone in a "bad mood" — throw open your heart and begin
to pour out unconditional love with great purpose of mind
and watch as darkness changes to light. You can LOVE a failing business back to life. You can LOVE a diseased body back
to wellness. And you can LOVE a negative condition right out
of existence!

Understand that the creative—creating Power of the Universe individualizes within your energy field and finds an outlet through your heart chakra *when you can love a person, place, thing or situation unconditionally.* The act (feeling) of unconditional love opens the chakra and propels the harmonizing energy directly into the low vibratory force field and begins to "perk up" the vibration—literally lifting it up to the Divine Standard. You are actually "shooting" Love Rays with this activity, and if you will just *practice* the procedure, you will be amazed at the results. But don't just think about it. Do it!

When you are not purposely using your Love-Power, continue to *be* the Presence of Love. For example, can the Presence of Love experience hurt feelings? Can Individualized Love feel rejected? Does a Master of Love attack out of anger? Would a Being of Love feel bitterness or resentment? Would God's perfect Expression of Love condemn or criticize? And could the Energy of Love ever be stagnant and static? You know the answers. Begin to live as the Love you are in Truth!

You might also spend some time in meditation reflecting on what Paul said about love in his letter to the Christians at Corinth:

"If I speak with the eloquence of men and of angels, but have no love, I become no more than blaring brass or crashing cymbals. If I have the gift of foretelling the future and hold in my mind not only all human knowledge but the very secrets of God, and if I also have that absolute faith which can move mountains, but have no love, I amount to nothing at all. If I dispose of all that I possess, yes, even if I give my own body to be burned, but have no love, I achieve nothing.

"This love of which I speak is slow to lose patience—it looks for a way of being constructive. It is not possessive: It is neither anxious to impress nor does it cherish inflated ideas of its own importance.

"Love has good manners and does not pursue selfish advantage. It is not touchy. It does not keep account of evil or gloat over the wickedness of other people. On the contrary, it is glad with all good men when truth prevails.

141

"Love knows no limit to its endurance, no end to its trust, no fading of its hope; it can outlast anything. It is, in fact, the one thing that still stands when all else has fallen."

CHAPTER SEVENTEEN

Your Role as the Christ

1. We must remember — through the awakening process — how to be the Masters we were created to be. We must understand the principle of supply so that we are not affected by anything that happens to the economic system. We must demonstrate radiant health so that we will have the energy and vitality to fulfill our purpose here. We must live under Divine Protection so that we may be in a safety zone at all times. We must reopen the Wisdom faculty within us so that we may be guided to take the right action at the right time and in the right way. And we must be a beacon of illumination for others seeking the spiritual path.

2. Too difficult a task for *you?* No, it really isn't. You can truly step out in mastery this very day *if you choose to do so!* I am not talking about strutting around like a peacock and calling yourself a Lord. I am referring to your acceptance of your true Identity this day, and letting that Identity live through you every single moment of your eternal life.

3. You see, Christhood is not something to come at a point in the future when you are more evolved. Christhood *is* — right now! I am the Christ of God. You are the Christ of God. We were *Christed* in the beginning, and nothing and no one can ever take that away from us. And while it is true that a

`part of us is asleep and under the spell of illusion, there is so much more of us that is fully awake, fully illumined and living the Reality of Truth—right now!

4. You are not having problems and you are not facing challenges. You never have and you never will. You are whole and complete. Abundance is yours, wellness is yours, loving and harmonious relationships are yours, total fulfillment is yours, and you are enjoying the exquisite perfection of God's Universe at this very moment. In Truth, you never left the Father's house, you never fell from grace, and you certainly were not tossed out of some garden and told to till the ground until you dropped dead.

5. Before you say that I have you mixed up with someone else, let me remind you that you are the offspring of the Infinite Mind and Power of the Universe, and all that this Infinite Mind and Power is, you are! For God's sake, *know this!* You finally got through the worm-of-the-dust fixation, but you are still hanging in there with the obscene notion that you are a pawn of fate fighting for your good in a hostile world.

6. The reason you are still playing the illusion game is that you are living out of the lower soul (ego) consciousness, but you don't have to. At any time in the last two thousand years you could have risen into the vibration of the Higher Soul and regained your realization of oneness with Spirit. In fact, you have touched this Christ Vibration in moments of meditation, but you did not lock into it. You took the elevator up, but instead of stepping out in mastery, you pushed the "down" button.

7. Regaining spiritual consciousness is not tearing one house down and building another. It is more like moving out of the basement and into the main floor. Remember that you are an individualized Energy Field. Within this force field are lower and higher vibrations. Your "I" of identity is like the mercury in a thermometer, rising and falling according to the weather of your thoughts and emotions. For the majority of people, the "I" remains in the cold, dark atmosphere of the lower mind, while all the time in the upper region of the

energy field, the Sun is shining and there is the warmth of Light, Love and Joy. Here, the Energy is pure, the vibrations high, and the Consciousness illumined.

8. But understand that this illumined Consciousness that is alive and living simultaneously with your mortal mind is *your* Consciousness. You cannot break a beam of light into parts. It's all one stream of radiance. When you lift up the "I" and move into this spiritual dimension of your own energy field, you do not lose your identity or your self-awareness. You do not lose consciousness and wake up as a stranger in a strange land. You do not trade one mind in for another. No, you take your consciousness with you and rise up into a new range of awareness, understanding and knowledge.

9. As your thinking and feeling natures move up to a different coordinate in your energy field, you *take on* the Consciousness of the Higher Soul and the dark energies of the lower self are transmuted. And while there were two states of consciousness before (duality), now there is only one. It is this Higher Soul Energy that says, "No one cometh unto the Father but by me." When you are in this spiritual vibration, you are literally "in tune with the Infinite" because the Infinite Spirit of God dwells in that vibration! In this Higher Consciousness you are fully aware — you understand — and you have Knowledge of the Living Christ of your Being.

10. As you rise into this finer vibration, you become aware of your God-Self as never before. You sense that intense Knowingness. You see the brilliance of the indwelling Light. You feel the fire of that boundless love within you. And as your mortal mind is totally swallowed up in this dimension of pure spiritual energy, you come into the fullness of the Christ Consciousness that says, "I am the Light of the world. I am the resurrection and the life." Now you *know* the Reality of your Self. Now you know that *you* are the Omnipotent Christ!

11. Will you begin the ascension today? Will you accept the Truth of your Divine Identity and begin to act the role of the Christ? Will you play the part? If you will, even though you may at this very moment be wallowing with the swine, I

assure you that the robe and ring will be yours sooner than you can possibly imagine. How do you play the role? With everything you've got!

12. You take the thought into your mind that you are the Christ of God and you *live* with that thought throughout the day. And from that parent thought will come mental children of great love, joy, faith, understanding, power, strength, wisdom, forgiveness, imagination, will, life, and enthusiasm. And if you momentarily "forget your lines"—ask yourself— how does the Christ think? And let the thoughts flow.

13. How does the Christ feel? With total and complete unconditional love . . . so begin this day to love everyone and everything as you have never loved before. *Feel* that love vibration. Let it flow, pour out, radiate from you. *Be* the Love of God in action! And if your emotions pause for a moment to be anxious and fearful, just remind yourself that those are not Christ-like feelings, and as the Christ of God, you will feel only joyous, peaceful, loving and happy emotions.

14. How does the Christ speak? With words that represent the attributes of God. So your words will be kind, loving, compassionate, fearless, wise, joyous. Not one word will come forth from your mouth that denies the Divinity of yourself or your fellow beings. Criticism is not a part of your nature; there is never a verbal attack. Right thinking precedes your words and your conversations are always uplifting and inspiring. Your voice is as music to ears eager for the message of Truth.

15. How does the Christ act? With actions that reflect illumination, love, power and perfect faith. Whenever there is an appearance of lack or limitation, you will give thanks for the infinite givingness of Spirit, and you will decree the Truth by recognizing the Reality behind the illusion. You will command the very Energy of your Being, the substance of your Self, to appear as the needed form or experience. Whenever you see conflict and hostility, you will radiate the Light of Love as a laser beam to harmonize the situation. Whenever there is the appearance of dis-ease, you will send forth the Light of Truth with intense radiation to dissolve the error pat-

tern and release the Healing Force from within the individual.

16. Remember, you are the Christ of God. Therefore, you will walk as the Christ, sit as the Christ, stand as the Christ. Your body language will reflect the Christ. Your facial expressions will reflect the Christ. And you will see everyone else as you see yourself . . . as the Christ of God . . . God in individual expression. Can you play the role? If you can, something very mystical and beautiful will happen to you. The "I" of identity will begin to rise, moving from the dark, cold atmosphere of mortal mind consciousness right up into the Kingdom of the Fourth Dimension. Even now you hear the call: "I am the bread of life. He that comes to me shall never hunger, and he that believes on me shall never thirst. You are from beneath; I am from above. Come unto me now."

17. I now return to the glory that I knew in the beginning. It is done!

APPENDIX

- Your Appointment to the Planetary Commission.

- The Time Where You Are at Noon Greenwich Time.

- The International Healing Meditation.

- Questions and Answers.

Please remove this page, date and sign it, note the country where you live, and mail to: The Quartus Foundation, P.O. Box 26683, Austin, Texas 78755.

I ACCEPT MY APPOINTMENT
TO THE PLANETARY COMMISSION

I choose to be a part of the Planetary Commission, and I do hereby consent to the healing and harmonizing of this planet and all forms of life hereon.

I shall begin this day to radiate the Infinite Spirit I AM in Truth to this world. I open my heart and I let Divine Love pour out to one and all, transmuting every negative situation and experience within the range of my consciousness.

I forgive everyone, including myself. I forgive the past and I close the door. From this moment on, I shall dedicate my life to turning within and seeking, finding, and knowing the only Presence, the only Power, the only Cause, and the only Activity of my eternal life. And I place my faith in the Presence of God within as my Spirit, my Substance, my Supply, and my Support.

I know that as I lift up my consciousness, I will be doing my part to cancel out the error of the race mind, heal the sense of separation, and restore the world to sanity.

With love in my heart, the thrill of victory in my mind, and joyous words on my lips, I agree to be a part of the world-wide group that will meet in spirit at noon Greenwich time on December 31, 1986, to release Light, Love and Spiritual Energy in the Healing Meditation for Planet Earth.

I now accept my appointment to the Planetary Commission!

Date	Signature	Country

This is your copy of your Appointment to the Planetary Commission. Leave it in the book for frequent review and rededication.

I ACCEPT MY APPOINTMENT
TO THE PLANETARY COMMISSION

I choose to be a part of the Planetary Commission, and I do hereby consent to the healing and harmonizing of this planet and all forms of life hereon.

I shall begin this day to radiate the Infinite Spirit I AM in Truth to this world. I open my heart and I let Divine Love pour out to one and all, transmuting every negative situation and experience within the range of my consciousness.

I forgive everyone, including myself. I forgive the past and I close the door. From this moment on, I shall dedicate my life to turning within and seeking, finding, and knowing the only Presence, the only Power, the only Cause, and the only Activity of my eternal life. And I place my faith in the Presence of God within as my Spirit, my Substance, my Supply, and my Support.

I know that as I lift up my consciousness, I will be doing my part to cancel out the error of the race mind, heal the sense of separation, and restore the world to sanity.

With love in my heart, the thrill of victory in my mind, and joyous words on my lips, I agree to be a part of the world-wide group that will meet in spirit at noon Greenwich time on December 31, 1986, to release Light, Love and Spiritual Energy in the Healing Meditation for Planet Earth.

I now accept my appointment to the Planetary Commission!

_____   _____   _____
Date   **Signature**   **Country**

TIME ZONES

The time where you are at noon Greenwich time.

United States Time Zones	Noon Greenwich Time
Pacific Standard Time	4:00 A.M.
Mountain Standard Time	5:00 A.M.
Central Standard Time	6:00 A.M.
Eastern Standard Time	7:00 A.M.

Noon Greenwich Time in Various Cities of the World

Berlin	1:00 P.M.	Montreal	7:00 A.M.
Buenos Aires	9:00 A.M.	Moscow	3:00 P.M.
Cairo	2:00 P.M.	Naples	1:00 P.M.
Copenhagen	1:00 P.M.	Nome	1:00 A.M.
Edmonton	5:00 A.M.	Ottawa	7:00 A.M.
Fairbanks	2:00 A.M.	Paris	1:00 P.M.
Glasgow	Noon	Rome	1:00 P.M.
Honolulu	2:00 A.M.	Sydney	10:00 P.M.
London	Noon	Tokyo	9:00 P.M.
Madrid	1:00 P.M.	Vancouver	4:00 A.M.
Mexico City	6:00 A.M.	Vienna	1:00 P.M.

INTERNATIONAL HEALING MEDITATION

In the beginning
In the beginning *God.*
In the beginning God created the heaven and the earth.
And God said Let there be light: and there was light.

Now is the time of the *new* beginning.
I am a co-creator with God, and it is a new Heaven
    that comes,
as the Good Will of God is expressed on Earth through me.
It is the Kingdom of Light, Love, Peace and Understanding.
And I am doing my part to reveal its Reality.

I begin with me.
I am a living Soul and the Spirit of God dwells in me, as me.
I and the Father are one, and all that the Father has is mine.
In Truth, I am the Christ of God.

What is true of me is true of everyone,
for God is all and all is God.
I see only the Spirit of God in every Soul.
And to every man, woman and child on Earth I say:
I love you, for you are me. You are my Holy Self.

I now open my heart,
and let the pure essence of Unconditional Love pour out.
I see it as a Golden Light radiating from the center of
    my being,
and I feel its Divine Vibration in and through me, above
    and below me.

I am one with the Light.

I am filled with the Light.
I am illumined by the Light.
I am the Light of the world.

With purpose of mind, I send forth the Light.
I let the radiance go before me to join the other Lights.
I know this is happening all over the world at this moment.
I see the merging Lights.
There is now one Light. We are the Light of the world.

The one Light of Love, Peace and Understanding is moving.
It flows across the face of the Earth,
touching and illuminating every soul in the shadow of
    the illusion.
And where there was darkness, there is now the Light
    of Reality.

And the Radiance grows, permeating, saturating every
    form of life.
There is only the vibration of one Perfect Life now.
All the kingdoms of the Earth respond,
and the Planet is alive with Light and Love.

There is total Oneness,
and in this Oneness we speak the Word.
Let the sense of separation be dissolved.
Let mankind be returned to Godkind.

Let peace come forth in every mind.
Let Love flow forth from every heart.
Let forgiveness reign in every soul.
Let understanding be the common bond.

And now from the Light of the world,
the One Presence and Power of the Universe responds.
The Activity of God is healing and harmonizing
    Planet Earth.
Omnipotence is made manifest.

I am seeing the salvation of the planet before my very eyes,
as all false beliefs and error patterns are dissolved.
The sense of separation is no more; the healing has
   taken place,
and the world is restored to sanity.

This is the beginning of Peace on Earth and Good Will
   toward all,
as Love flows forth from every heart,
forgiveness reigns in every soul,
and all hearts and minds are one in perfect understanding.

It is done. And it is so.

Questions and Answers

1. Q. Why do we have to wait until 1986 to start healing the world? Who knows what might happen between now and then!

A. You don't have to wait. Remember that you help the world when you lift up your consciousness and become one with the Christ Vibration within. So begin this day to clean up, clear out and expand your consciousness, knowing that *you* can make a difference right now! The world-wide Quartus membership is using the International Healing Meditation daily at 9:00 AM and 10:00 PM central time, and your participation now will release even more Light for the harmonizing of the planet.

We feel that it will take the 2-plus years between the time this book is published and December 31, 1986, to create the proper awareness of the Planetary Healing Day—and it's going to take a true grass roots effort to reach the millions of people who will consent to participate. Perhaps you will consider the idea of collecting signatures from friends and relatives, and asking them to do likewise. This kind of personal activity has been going on since early 1984 in the United States, Canada, England, and Europe, and by now I am sure that it has spread to other countries. Remember that we do not want addresses—just the name of the individual and the country. The signatures should be sent to P.O. Box 26683, Austin, Texas 78755.

2. Q. Does the idea of "Superbeings" or "Masters" contribute to spiritual snobbery among New Age people?

A. If it does, it's that bloated ego getting in the way again. Please understand this: Every soul on this planet, whether a New Thoughter, Mainliner, Fundamentalist, Catholic, Jew, Buddhist, Hindu, or whatever, is a master. No one is more spiritually advanced than another. The only difference is that some have awakened to their true Identity. The 12 phases of consciousness that I wrote about in *The Superbeings* are simply that— phases in the awakening process. Some of our brothers and sisters are asleep; they are sleeping masters. Some are awake; they are the awakened masters. And millions are just now coming out of the deep sleep. Let's call them drowsy masters. As you work with this book to eliminate false beliefs, error patterns and negative appearances, you will experience a rise in consciousness. And as you develop a greater awareness, understanding and knowledge of the Christ within, your outer world will change accordingly. But even when total dominion comes, you will not be more advanced than your neighbor. In fact, the greater the spiritual evolvement, the greater the desire to *serve* . . . the higher you rise in consciousness, the more you will express unconditional love toward one and all. An Awakened One has no false pride or spiritual arrogance!

3. Q. I have heard that there has been some opposition to the New Thought Movement because it believes in the divinity of man. Is this true?

A. There are some groups who continue to cling to the absurd idea that man is a miserable sinner and worm of the dust. Yet most profess to believe in Jesus Christ. If we look at the words and teachings of Jesus in the New Testament, we'll see that his primary purpose for coming into this world was to reintroduce the concept of the innate perfection of man. Understand that Jesus would never tell us to do something that would be considered impossible, so when he said "Be ye therefore perfect, even as your Father which is in heaven is perfect" (Matt. 5:48)—he was telling us to awaken to the Truth of our Identity. Jesus also stated that "the kingdom of God is within you" (Luke 17:21)—not on the other side of some distant cloud. And he taught the oneness of man and God when he prayed, "That they all may be one; as thou, Father, art in me, and I in thee, that they also may be one in us . . ." (John 17:21). He also firmly established the Father-son relationship of God and man when he said, "And call no man your father upon the earth: for one is your Father, which is in heaven" (Matt. 23:9). He identified himself as the light of the world— and then said "Ye are the light of the world" (Matt. 5:14). In other words, our true nature is identical to Jesus. And through Christ indwelling ". . . nothing shall be impossible unto you." Would a worm of the dust have dominion over this world? No! But a Son of God does—and that's exactly Who and What you are! Jesus confirmed this in John 10:34 when he said "Ye are gods." Continue on through the Bible and you'll see that the primary teaching is on the Oneness of God and man. "The Holy Spirit which is *in* you . . ." (I Cor. 6:19). ". . . it is the same God which worketh all *in* all" (I Cor. 12:6). ". . . stir up the gift of God, which is *in* thee." (II Tim. 1:6). (The italics are mine.)

4. Q. How do you define the antichrist?

A. Any individual or group who denies the divinity of man as exemplified by Jesus Christ, i.e. to be in opposition to "Christ in you"—the indwelling Christ or Higher Self of each individual. The ego or mortal mind may also deceive you, and is often considered the voice of the antichrist, however, the illumined ones have identified this inner adversary as Satan or Lucifer—the "personality" phase of mind that tempts man to believe he is separate and apart from God.

5. Q. Are New Thought teachings and Ancient Wisdom the same thing?

A. The common denominator of metaphysics is there, but I would consider New Thought more as spiritual psychology and Ancient Wisdom more as esoteric philosophy. Another way to look at it would be to say that New Thought is Ancient Wisdom Westernized, Christianized, and Pragmaticized. In the coming years the New Thought teachings will place greater emphasis on energy, vibrations, centers of force, the rays, etc., while the Esotericists will focus more on the *one* Master within each soul.

6. Q. How can I convince my husband that reincarnation is true?

A. You can't. When he is ready to change his belief system and accept new ideas, he will, but to try to force your beliefs on him will only damage your relationship. Based on the 100th Monkey concept (when enough people believe something is true, the idea reaches critical mass and becomes true for everyone), the acceptance of reincarnation will one day be universal. It is interesting that throughout our recorded history, reincarnation was a part of the human belief system. In fact, for more than 500 years after the beginning of Christianity, the concept of reincarnation was part of the Christian teaching. But in A.D. 553, at the second Ecclesiastical council of Constantinople, it was banished from "official" Christianity. Here was a doctrine that had been a teaching of Buddha and accepted by Jesus . . . one that had been taught by the Hindus, the Egyptians, the Greeks, and the Jews. Yet it was dismissed with the stroke of a pen by a council of men. Fortunately, the majority of people on the planet still believe that "reincarnation is a fact."

7. Q. Are the New Age Children supposed to be "different" or more spiritually minded?

A. We have been told that beginning in 1976, the majority of babies born into the earth plane have been illumined ones. And in our travels around the country we have seen and heard evidence of this. Examples would include the three year old boy in Oklahoma who healed his mother when she suffered a stroke . . . a four year old girl in New York who began teaching her parents metaphysical truths during a period of family friction . . . a large number of boys and girls who can read and comprehend what they are reading by the age of two . . . little ones who see auras, who know what you're going to say before you speak, who radiate Light and great spiritual energy, and who know the meaning of love better than most adults.

Marcus Bach has written that "there is a unique breed of modern children endowed with an innate tendency for love, gifted with super-normal intuition, and equipped with a vision of heightened spiritual values and a more peaceable world." As parents, you have a definite responsibility to assist your children in fully awakening and understanding their mission and purpose. And you can best do this in a home permeated with peace and love — and by practicing what you teach. These children have psychic introspections and the ability to mature much more rapidly than the "average" child, and they can read you like a book. Here are a few suggestions:

- Develop an atmosphere in the home of trust, love and humor. As Marilyn Ferguson writes in *The Aquarian Conspiracy,* "One key is authenticity . . . parents acting as people, not as roles."
- Before they enter the educational system, take a page

from Socrates. He defined education as withdrawing wisdom from the mind of the child out to where he could examine and use that which was already his from the very beginning. He proved this by working with the slave boy in Meno, showing that by skillful questioning, the boy already knew geometry. He simply helped the boy to *remember* something that was not a part of his objective consciousness.

- Consider your children as *individuals*—and treat them as you would want to be treated.
- Eliminate the words "bad" and "afraid" from your vocabulary.
- Don't constantly "schedule" your child. Allow time for him or her to "Be."
- Take time to "be still" with your child.
- Stimulate imagination and self-expression.
- Develop an understanding of our kinship with all life—all nature—the rocks, plants, animals
- Do not consider your children as your possessions. You don't "own" them. You are simply a caretaker. And they are not to be molded in your image because they are already created in the image of God!
- Daily acknowledge the Christ in your child. See him/her as whole, strong, victorious, bright, joyous, flawless, fearless and free . . . an individualization of God on earth.
- Listen more and talk less.
- Be fair.
- Be honest.
- Love each one unconditionally, with absolutely no strings attached.

8. **Q. I don't know if I can believe everything I read in all the New Age books. Do you?**

A. I try to be open-minded regarding everything I read or hear while at the same time practicing spiritual discernment. The research at Quartus covers a broad area of metaphysical-esoteric topics and teachings, some of which are beyond my third-dimensional level of understanding. That's why I take everything into the inner chamber of Light for greater comprehension and appreciation. And sometimes I "see" the red flag signifying that the particular message is coming from a lower vibration consciousness, meaning that I should move on to something else. Each one of us must do this . . . we must sharpen our intuition so that we do not waste our time . . . so that we are not led astray. Just remember that all Truth must come from the Spirit of Truth within!

9. **Q. Are we being directed by Spiritual Masters from another realm?**

A. The "Pool of Wisdom" from the Illumined Ones is being drawn into the consciousness of the seekers as guidance. The sum of the whole is serving the parts until the Light from within each individual may shine forth as spiritual illumination.

10. **Q. Do you have evidence that meditation and spiritual "treatment" can really make a difference in a person's life?**

A. The Quartus files are filled with letters from men and women who have moved into new dimensions of consciousness through daily spiritual work, and who are enjoying the fruits of this higher vibration. A few examples follow.

- "Miracle follows miracle! Something which I timidly whispered to God about, never daring to come right out and ask for it, has come to pass. And the circumstances are so remarkable that it's clear that this miracle was only waiting for me to be ready to receive it. Now I know that all things are possible to he who believes."

- "Since I have become aware of the One Presence and One Power and became a partner with Same, my life is a daily miracle. Of course, one has to be aware, and constantly aware. I do feel blessed to KNOW 'all things are possible to him who believes' . . . 'as a man thinketh in his heart, so is he' . . . and 'all things be ready if the mind be so'. The above to me aren't just quotes but TRUTH, which I prove to myself everyday."

- "I am producing results so fast and with so much abundance that at times I am frightened. I feel like a laser gun from some 'Star Trek' episode, the difference being that it's not laser light beams but power and creative energy flowing through me out into the world. I'm producing results which the physical world attests to."

- "For the past two years I've felt like a sponge, absorbing all I can read. Sometimes the absorbing has been pleasant, sometimes not. But I realize that the boundaries are set by me personally. When I rush to learn more, it's almost more than I can personally comprehend, so I always go back to step one, which is to allow time to comprehend, then I'm ready to learn more. When I personally came to the end of my rope and decided to let go and let God, I found the Kingdom within. The knowledge that came with that was mind boggling. Suddenly all the pieces fit together."

- "I finally learned the lesson of 'I am the good that you seek'. I was seeking the result and thus concentrating on my needs rather than the Presence and the knowledge that *I* am the Cause. I feel so good! The miracles began happening yesterday and they are piling up one after another now."

- "Late one night as I rounded a curve on my way

home, a car in the opposite lane traveling 100 miles per hour could not make the curve, slid in front of me, hit the curb, turned upside down and flew over the hood of my car. I felt protected throughout . . . no one was seriously hurt in the other car and all was well. I felt wrapped in a warm cocoon of love and life, and the feeling stayed very intense for days. During that time the universe continued to shower me with blessing upon blessing."

- "I just wanted you to know how much we love and bless our new home. It was absolutely amazing that once I opened up my channels for receiving, the money just manifested for the down payment . . . $13,000 in 48 hours. God works wonders in the most mysterious ways!"

- "I have just completed the 40 day program for prosperity. It works! I needed new furnishings for my bedroom — what I had was all worn out and I did not know where the money was coming from to make the changes. Then the day after I completed the program I received some money, but the amount could not buy all I needed because furniture in this area is expensive. So I said, 'Dear Lord, this is what I have . . . I'm going out there with it so direct me to the right place where I can get what I want for what I have'. Well, I found the right furnishings — just as I had visualized — for the right price with $100 left over and delivered the same day. God is so wonderful!"

- "I went to the ballet on Sunday afternoon. The parking garage was dark, so I turned on the car lights. No spaces. I ended up on the roof in the bright sun, forgot my lights and left them on. Three hours later, on my return, all the battery would say was one tiny 'uhh' when I turned the key. I prayed and tried it again several times. 'Uhh' — nothing more. I acknowledged my error, asked forgiveness for it, and forgiveness for my belief that a dead battery meant the car wouldn't go; acknowledged that God is the *only* power, that the appearance of the dead battery was an appearance of illusion only, that the past was over and could touch me not, that there is no order of difficulty in miracles, that Holy Spirit was solving the problem for me this very instant, and that I knew that all the Love in the universe was pouring energy into my battery at this very instant. That took one and a half minutes. I turned the key again, and the motor started right up, no sign of further difficulty. The postscript is that I told my assistant about it the next day. On Tuesday she went to the store in a driving rainstorm, turned on her lights, forgot them, came back to a 'dead' battery that would say nothing more than 'uhh'. She meditated and 'remembered' for a minute and a half, and *her* battery flared right up, no further trouble. She acknowledged some doubt, but she believed enough to do it anyhow, and so it worked."

- ". . . I am so happy to tell you that everything has turned around and that my world is now filled more than ever with light, peace and happiness. When I last wrote I was overcome with such fear and loneliness — and I was out of work to boot. But when I let go and let God, my whole world became filled with light. I literally fell into a job that I love, and several new clients came my way. My prosperity is growing and surely I am on my way to financial independence. But more than that, all fear has completely disappeared, and I KNOW that God is ever my Source."

- "Now the healing energy flows and heals through no effort of my own, other than to be what I AM, have always been now and forever — LOVE. What Jesus did, we all can do. I know. I have. If I can, all can — by becoming aware of who they are and what they are and claiming their divine inheritance."

11. Q. What is the most effective form of meditation and spiritual treatment?

A. The one that works for you. Remember that your good already IS! You already have in the depths of your consciousness the spiritual prototype of everything you could possibly desire in the physical world. Your role in the scheme of things is to clear your consciousness and become an open channel for the outpouring of your good. And the "way" to do this depends on your particular mind-set and vibration of consciousness. As you spend time with the "lessons" in the workbook, and devote a period each day to contemplative meditation, you will open the door to the Secret Place. The Spirit within you guarantees it!

12. Q. Can we evolve in consciousness more rapidly through group activities?

A. If you mean *spiritual* groups, the answer is definitely yes. The time of spiritual growth through total isolation is gone. It went out with the Age of Pisces. In the Aquarian Age, the Age of Spirituality, the emphasis is on two or more gathered together in the name and through the power of the indwelling Christ. This is why it is so important to get involved with a New Thought church, study group, or a weekly class where experiences are shared. When the group is on the same vibration, spiritual treatments and "master minding" are very powerful, frequently overcoming the limitations of time and space in producing new conditions. And the sharing of energies can definitely enhance spiritual growth.

ABOUT THE AUTHOR

John Randolph Price is the president and executive director of The Quartus Foundation for Spiritual Research, Inc., a non-profit organization headquartered in the Texas hill country near Austin. His wife, Jan, co-founder of Quartus, chairs the advisory board for the Foundation — and together they conduct seminars across the country based on the great spiritual awakening that is taking place on Planet Earth.

They have appeared on radio and television in major cities; have presented workshops in Divine Science, Religious Science, Unity, and independent New Thought churches, and have participated as guest lecturers in symposiums sponsored by the International New Thought Alliance, the Association for Research and Enlightenment, and other metaphysical groups. Their backgrounds in spiritual research and the application of metaphysical principles span more than 15 years.

In addition to *The Planetary Commission*, John has authored *The Superbeings*, and *The Manifestation Process: 10 Steps to the Fulfillment of Your Desires*.

The Quartus Foundation

The Quartus Foundation for Spiritual Research is an organization dedicated to research and communications on the divinity of man. We seek to study the records of the past, investigate events and experiences of the present, and probe the possibilities and potential of the future through the illumined consciousness of awakened Souls.

Our objective is to continually document the truth that man is a spiritual being possessing all of the powers of the spiritual realm . . . that man is indeed God individualized, and that as man realizes his true identity, he becomes a Master Mind with dominion over the material world.

The documentation comes through indepth research into case histories of the past and present, which reveal the healing, prospering, harmonizing Power of God working in and through man. We believe that man is capable of rising above every problem and challenge that could possibly beset him, and that he is doing this daily in ways that are considered both "mysterious and miraculous." But in truth, the evil, the illness, the failure, the limitation, the danger, the injustice disappear through a change in individual consciousness. What brought about the change? We seek to examine closely the problem and the solution, the activity of Mind and the Law of Mind, the cause and effect—and to build a fund of knowledge based on the interrelationship of Spirit, Soul, Body, and the world of form and experience.

We believe that what one person is doing to alter conditions and reveal order and harmony, all can do—and by researching and communicating specific examples of man's inherent powers, we can do our part in assisting in the general upliftment of consciousness. Through greater understanding, we can all develop a more dynamic faith in our inner Self, a conviction that our potential is only limited by the scope of our vision, and a Knowledge that mankind is Godkind and does not have to accept anything less than heaven on earth.

For complete details on the activities of The Quartus Foundation, write for your free copy of The Quartus Report, P.O. Box 26683, Austin, Texas 78755.

BOOKS AND TAPES AVAILABLE FROM THE QUARTUS FOUNDATION

Books

- *The Planetary Commission*
  by John Randolph Price $7.95

- *The Superbeings*
  by John Randolph Price $5.95
  Men and women in all walks of life are rapidly evolving
  toward undreamed of powers . . . some have reached
  the point of mastery where they are no longer bound
  by the ills, limitations and problems of this world.
  Their secrets are revealed in this exciting book so that
  you, too, may develop and use the miracle power of
  the Supermind.

- *The Manifestation Process: 10 Steps to the*
  Fulfillment of Your Desires
  by John Randolph Price $3.95
  This booklet is an expanded version of the process
  taught in the early Superbeing Seminars, and is based
  on a concept originally developed after closely evaluat-
  ing the consciousness characteristics of a number of
  highly evolved individuals. Each principle is discussed,
  and the reader is led through a meditative treatment to
  duplicate the automatic activity of Superbeing Consciousness in a step-by-step process.

- *Journey of Love* by Alan Mesher $6.95
  A combination of personal experiences, spiritual teachings and meditative practices that will help the reader
  develop and bring forth the power, fulfillment and
  mastery from within.

Cassette Tapes

By John Price

- *2The 40-day Prosperity Plan —*
  including the technique of how to
  listen to your money talk $6.95

- *The Future Is Now —* from Chapter One of *The Plane-*
  tary Commission — Discusses the critical opportunity
  before us to heal and harmonize the planet, and
  includes the Healing Meditation $6.95

- *The Incredible Power of Love* $6.95

- *Developing Mastery Through Faith,*
  Strength and Wisdom $6.95

- *The Manifestation Process —*
  companion to the booklet
  with the Manifestation Meditation $9.95

By Jan Price

- *Releasing and Infilling Meditations* $5.95

- *Have a Love Affair with Your Self* $6.95

- *The Miracle Working Power of At-One-Ment* $6.95

Please add $1.00 for postage and handling. Texas resi-
dents add 5% Sales Tax.
Order from: The Quartus Foundation, P.O. Box 26683,
Austin, Texas 78755